The Cambridge Manuals of Science and Literature

PREHISTORIC MAN

PREHISTORIC MAN

BY

W. L. H. DUCKWORTH
M.A., M.D., Sc.D.

University Lecturer in Physical
Anthropology, Cambridge

Cambridge:
at the University Press
1912

CAMBRIDGE UNIVERSITY PRESS
Cambridge, New York, Melbourne, Madrid, Cape Town,
Singapore, São Paulo, Delhi, Tokyo, Mexico City

Cambridge University Press
The Edinburgh Building, Cambridge CB2 8RU, UK

Published in the United States of America by Cambridge University Press, New York

www.cambridge.org
Information on this title: www.cambridge.org/9781107606012

© Cambridge University Press 1912

First published 1912
Second Edition, 1912
First paperback edition 2011

A catalogue record for this publication is available from the British library

ISBN 978-1-107-60601-2 Paperback

Cambridge University Press has no responsibility for the persistence or accuracy of URLs for external or third-party internet websites referred to in this publication, and does not guarantee that any content on such websites is, or will remain, accurate or appropriate.

With the exception of the coat of arms at the foot, the design on the title page is a reproduction of one used by the earliest known Cambridge printer, John Siberch, **1521**

PREFACE

THIS book deals with the earliest phases in the past history of Mankind: the selected period ends at the Aurignacian division of the Palaeolithic Age. I regret to be unable to affix definite dates in years to the several divisions of time now recognised. To illustrate the difficulty of forming conclusions on this subject, it should be noted that in 1904 Professor Rutot (p. 103) assigned a duration of 139,000 years to the Pleistocene period, while in 1909 Dr Sturge claimed 700,000 years for a portion only of the same period. Evidently the present tendency is to increase enormously the drafts on geological time, and to measure in millions the years that have elapsed since the first traces of human existence were deposited.

But in the face of estimates which differ so widely, it seemed preferable to distinguish subdivisions of time by reference to animal-types or the forms of stone-implements, rather than by the lapse of years.

In the attempt to summarise a considerable amount of evidence, I have tried to select the facts most relevant to the subject in hand. And where an opinion is expressed I have endeavoured to indicate the reasons for the decision that is adopted.

Additional evidence is pouring in at the present time, and there is no doubt but that the next few years will witness great extensions of knowledge. In this connection, I take the opportunity of mentioning the discovery made a few weeks ago by M. Henri Martin at La Quina, of a human skeleton resembling the Neanderthal type but presenting (it is said) definite features of inferiority to that type. Another subject of vast importance is Mr Moir's recent demonstration (p. 106) of elaborately worked implements resting beneath strata referred to the Pliocene period.

For the loan of blocks, or for permission to reproduce illustrations, my cordial thanks are due to the editors and publishers of the journals mentioned in the following list. The authors' names are appended to the several illustrations.

Anatomischer Anzeiger,
Archiv für Anthropologie,
Archivio per l'Antropologia e la Etnologia,
Beiträge zur Urgeschichte Bayerns,
Korrespondenzblatt der deutschen anthropologischen Gesellschaft,
L'Anthropologie,
Royal Dublin Society,
Royal Society of Edinburgh,
Zeitschrift für Ethnologie.

W. L. H. DUCKWORTH

December 11, 1911

CONTENTS

CHAP.		PAGE
I.	The Precursors of Palaeolithic Man	1
II.	Palaeolithic Man	17
III.	Alluvial Deposits and Caves	63
IV.	Associated Animals and Implements	85
V.	Human Fossils and Geological Chronology	112
VI.	Human Evolution in the light of recent research	127
	Table A . . . . *to face p.* 85	
	„ B . . . . . . *to* „ „ 118	

LIST OF ILLUSTRATIONS

FIG.		PAGE
1.	Outline tracings of skulls of Pithecanthropus etc. (From Dubois)	5
2.	Outline tracings of Jawbones, (A) Mauer (B) ancient Briton	11
3.	Tooth from Taubach: surface of crown. (From Nehring)	22
4.	Tooth of Chimpanzee. (From Nehring)	22
5, 6.	Tooth from Taubach: inner and outer sides. (From Nehring)	23
7.	Human skull from Krapina. (From Birkner)	25
8.	Tracings of teeth from Krapina and Mauer. (From Kramberger)	29
9.	Human skull from La Chapelle-aux-Saints. (From Birkner)	33

LIST OF ILLUSTRATIONS

FIG.		PAGE
10.	Outline tracings of skull from La Chapelle-aux-Saints etc. (From Boule)	35
11.	Contours of skulls, (A) New Guinea man (B) European woman	36
12.	Outline tracing of human skull from Le Moustier	40
13.	Outline tracings of jawbones from Mauer and Le Moustier	41
14.	Outline tracings of jawbones from Mauer, La Naulette, etc. (From Frizzi)	42
15.	Outline tracings of jawbones, (A) ancient Briton (B) Le Moustier (C) Mauer	43
16.	Outline tracings of the Forbes Quarry (Gibraltar) skull. (From Sera)	48
17.	Human skull of the Grimaldi-type. (From Birkner)	51
18.	Outline tracings of skulls from Galley Hill etc. (From Klaatsch)	58
19.	Section of the strata at Trinil in Java. (From Dubois)	64
20.	View of the Mauer sand-pit. (From Birkner)	65
21.	Section of the Krapina rock-shelter. (From Birkner)	69
22.	Plan of the cave at La Chapelle-aux-Saints. (From Boule)	72
23.	Two sections of the Grotte des Enfants, Mentone. (From Boule)	77
24.	Chart of the relative duration of Miocene, Pliocene, and Pleistocene time. (From Penck)	107
25.	Chart of oscillations of snow-level in the Glacial period. (From Penck)	119
26.	Outline tracings of skulls of Pithecanthropus etc. (From Dubois)	129
27.	Position of Palaeolithic Man in the scale of evolution. (From Cross)	131
28.	Thigh-bones arranged to illustrate Klaatsch's theory	136
29.	The human skeleton found beneath the Boulder-clay at Ipswich. (From a drawing by Dr Keith, reproduced with permission)	153

CHAPTER I

THE PRECURSORS OF PALAEOLITHIC MAN

Our knowledge of prehistoric man is based naturally upon the study of certain parts of the human skeleton preserved in a fossil state. In addition to these materials, other evidence is available in the form of certain products of human industry. These include such objects as implements of various kinds, owing their preservation to the almost indestructible nature of their material, or again artistic representations, whether pictorial or glyptic.

The evidence of the bones themselves will be considered first, partly for convenience and partly in view of the cogency possessed by actual remains of the human frame. Other branches of the subject will come under review afterwards.

Of all the discoveries of ancient remains, whether possibly or certainly human, two in particular stand out pre-eminently in marked relief. The specimens thus distinguished are known as the remains of *Pithecanthropus erectus*, on the one hand, and on the other a jaw-bone which is attributed to a human type described (from the locality of the discovery) as *Homo heidelbergensis*.

The geological antiquity assigned in each instance

is greater than that claimed for any bones acknowledged unreservedly to be human.

It is thus clear that a high value attaches to these specimens if they be regarded as documents testifying to the course of human evolution. When the bones are examined, the contrast they provide with all human remains is so marked as to emphasise at once the necessity for a thorough and critical examination of their structure.

Pithecanthropus erectus.

In the case of these bones, the facts are now so widely known and so easily accessible as to render unnecessary any detailed exposition here. The discoveries were made in the years 1891 and 1892 by Professor Dubois[1], who was engaged at the time on an investigation of the remains of various animals found embedded in a river-bank in Java. As is well known, the actual remains are scanty. They comprise the upper part of a skull, part of a lower jaw (which has never been described), three teeth, and a left thigh-bone.

Before entering upon any criticism of the results of Professor Dubois' studies, it is convenient to give a general statement of his conclusions. Here we find described a creature of Pliocene age, presenting a form so extraordinary as hardly to be considered

[1] The numbers refer to the Bibliography at the end of the volume.

human, placed so it seems between the human and simian tribes. It is Caliban, a missing link,—in fact a Pithecanthropus.

With the erect attitude and a stature surpassing that of many modern men were combined the heavy brows and narrow forehead of a flattened skull, containing little more than half the weight of brain possessed by an average European. The molar teeth were large with stout and divergent roots.

The arguments founded upon the joint consideration of the length of the thigh-bone and the capacity of the skull are of the highest interest. For the former dimension provides a means of estimating approximately the body-weight, while the capacity gives an indication of the brain-weight. The body-weight is asserted to have been about 70 kgm. (eleven stone) and the brain-weight about 750 gm. And the ratio of the two weights is approximately $\frac{1}{94}$. The corresponding ratios for a large anthropoid ape (Orang-utan) and for man are given in the table following, thus:

Orang-utan	$\frac{1}{183}$
Pithecanthropus erectus	$\frac{1}{94}$
Man	$\frac{1}{51}$

The intermediate position of the Javanese fossil is clearly revealed.

The same sequence is shewn by a series of tracings representative of the cranial arc in the middle line of the head (Fig. 1). And the results of many tests of this kind, applied not only by Professor Dubois but also by Professor Schwalbe, are confirmatory of the 'intermediate' position claimed for *Pithecanthropus erectus*. The molar teeth are of inadequate size if the skull-cap is that of an ape, whereas they are slightly larger than the corresponding teeth furnished by primitive existing human types. And now some of the objections to this account may be taken.

In the first place, the claim to Pliocene antiquity is contested. So keen an interest was excited by Professor Dubois' discovery that more than one expedition has been dispatched to survey and review the ground. It is now declared in certain quarters that the horizon is lower Quaternary : I do not know that any attempt has been made to reduce the age of the strata further. As the matter stands, the difference is not very material, but Professor Dubois refuses to accept the revised estimate and still adheres to his own determination. Incidentally the more recent work (Blanckenhorn[2], 1910) has resulted in the discovery of a tooth claimed as definitely human (this is not the case with the teeth of *Pithecanthropus erectus*), and yet of an antiquity surpassing that of the remains found by Professor Dubois. The latter appears unconvinced as to the genuineness of the find, but no doubt the case will be fully discussed in publications now in the course of preparation.

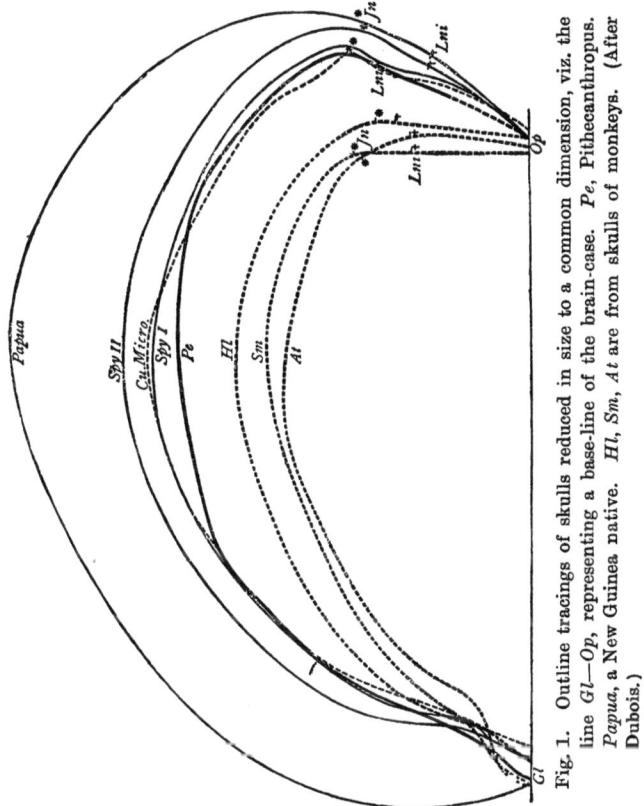

Fig. 1. Outline tracings of skulls reduced in size to a common dimension, viz. the line *Gl—Op*, representing a base-line of the brain-case. *Pe*, Pithecanthropus. *Papua*, a New Guinea native. *Hl*, *Sm*, *At* are from skulls of monkeys. (After Dubois.)

Professor Dubois assigned the bones to one and the same skeleton, and for this he has been severely criticised. Apart from arguments affecting the geological age of the specimens, the question of their forming part of a single individual is very momentous. For if two skeletons are represented, one may be human, while the other is that of an ape. It is admitted that the larger bones were separated by a distance of forty-six feet. By way of meeting this criticism, it is submitted that the distance is by no means so great as to preclude the possibility of the common and identical origin of the various bones. Moreover it is at least curious that if two skeletons are here represented, no further remains should have been detected in the immediate vicinity.

The fact that the thigh-bone might easily have passed as that of a man, while the skull-fragment is so divergent from all modern forms as to be scarcely human, is of great interest. The contrast between the indications provided by the two bones was remarked at once. Some writers, rejecting certain other evidence on the point, then drew the inference that the human thigh-bone had been evolved and had arrived at the distinctive human condition in advance of the skull. The importance of this conclusion lies in the fact that the human thigh-bone bears indications of an erect attitude, while the form of the skull gives guidance as to the size of the brain, and

PRECURSORS OF PALAEOLITHIC MAN

consequently to some extent provides a clue to the mental endowment of the individual. Whether the erect attitude or the characteristic brain-development was first obtained by man has been debated for many years. In this case, the evidence was taken to shew that the assumption of the erect attitude came as a means of surmounting the crux of the situation. Thenceforth the upper limb was emancipated entirely from its locomotor functions. Upon this emancipation followed the liberation of jaws and mouth from their use as organs of prehension. Simultaneously the mechanism whereby the head is attached to the neck and trunk became profoundly modified. This alteration gave to the brain an opportunity of growth and increase previously denied, but now seized, with the consequent accession of intellectual activity so characteristic of the Hominidae.

The story thus expounded is attractive from several points of view. But while possessing the support of the Javan fossil remains, it is not confirmed in the embryonic history of Man, for there the growth of the brain is by far the most distinctive feature. Nor did those who adopted this opinion (in 1896), take into account all the characters of the ancient human remains even then available. For the evidence of those remains points to an order exactly the reverse of that just stated, and it indicates the early acquisition of a large and presumably active brain. And now that additions have been lately made to those older remains (other than the Javan bones), the same 'reversed' order seems to be confirmed. On the whole therefore, the soundest conclusion is that following a preliminary increment of brain-material, the erect attitude came as a further evolutionary advance.

But to return from this digression to the objections against the *Pithecanthropus erectus*, it must now be explained that the very contrast between the thigh-bone and the skull-cap in respect of these inferences, has been used as an argument against the association of these bones as part of one skeleton.

The objection may be met in two ways at least. For instance, the thigh-bone may yet possess characters which lessen its resemblance to those of recent men, but are not recognised on a superficial inspection. Careful investigation of the thigh-bone seems to shew that such indeed is the case (indeed the human characters are by some absolutely denied). But together with this result comes the discovery that the characters of straightness and slenderness in the shaft of the bone from which the inference as to the erect attitude was largely drawn, do not give trustworthy evidence upon this point. In fact, a human thigh-bone may be much less straight and less slender than that of arboreal animals such as the Gibbon, the Cebus monkey, or the Lemurs (especially Nycticebus). The

PRECURSORS OF PALAEOLITHIC MAN

famous Eppelsheim femur is straighter than, and as slender as that of Pithecanthropus. It was regarded at first as that of a young woman, but is now ascribed to an anthropoid ape. And in fact, even if the skull-cap and thigh-bone of Pithecanthropus should be retained in association, it seems that the title 'erectus' is not fully justified.

Another method of rebutting the objection is based on the suggestion that Pithecanthropus is not a human ancestor in the direct line. Thus to describe an uncle as a parent is an error not uncommon in palaeontology, and it was treated leniently by Huxley. To my mind this position can be adopted without materially depreciating the value of the evidence yielded by the conjoint remains, provided only that their original association be acknowledged. Should this assumption be granted, the claims put forward on behalf of his discovery by Professor Dubois seem to be justified, On the other hand, should the association of skull-cap and thigh-bone be rejected, the former has not lost all claim to the same position. For the most recent researches of Professor Schwalbe[3] of Strassburg, and the further elaboration of these by Professor Berry[4] and Mr Cross[5] of Melbourne, support Professor Dubois' view. And though the objections may not have been finally disposed of, a review of the literature called forth by Professor Dubois' publications will shew a slight margin of evidence for, rather than against his view.

The Heidelberg or Mauer Jaw[6].

Professor Dubois' Javanese researches were carried out in the years 1891 and 1892. Fifteen years separate the discovery of the *Pithecanthropus erectus* from that of the second great find mentioned in the introductory paragraph of this chapter. This period was by no means barren in respect of other additions to the list of human fossils. But the other results (including even the finds at Taubach) are regarded as of subsidiary importance, so that their consideration will be deferred for the present. In 1907 a lower jaw, known now as the Heidelberg or Mauer jaw, was discovered by workmen in the sand-pit of Mauer near Heidelberg.

The Mauer jaw is indeed a most remarkable specimen. The first general outcome of an inspection of the photographs or of the excellent casts (which may now be seen in many museums) is a profound impression of its enormous strength (Figs. 2, 13, and 15 *c*). By every part of the specimen save one, this impression is confirmed. This massiveness, together with the complete absence of any prominence at the chin, would have caused great hesitation in regard to the pronouncement of a decision as to the probable nature of the fossil. The one paradoxical feature is the relatively small size of the teeth. All of these have been preserved, though

I] PRECURSORS OF PALAEOLITHIC MAN 11

on the left side the crowns of four have been removed by accident in the process of clearing away some adherent earth and pebbles. The net result shews

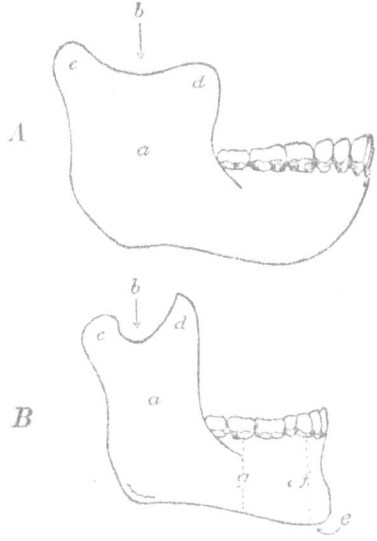

Fig. 2. *A* outline tracing of a cast of the Mauer Jawbone. *B* a similar tracing from an unusually large jaw of an ancient Briton. (From specimens in the Cambridge Museum.)

that the teeth are actually within the range of variation provided by human beings of races still extant, though commonly regarded as 'primitive,' if not pithecoid (such as the aboriginal race of Australia). Yet these teeth are implanted in a jaw of such size and strength as render difficult the reference of the specimen to a human being.

The most striking features of the Mauer jaw have been mentioned already. Before entering upon a further discussion of its probable nature, it will be well to note some of the other distinctive characters. Thus the portion Fig. 2 (*a*) known technically as the ascending ramus is of great size, and particularly wide, surpassing all known human specimens in this respect. The upper margin of this part is very slightly excavated, a slight depression (*b*) replacing the very definite 'sigmoid' notch found in almost all human jaws (though the relative shallowness of this notch has been long recognised as distinctive of the lowest human types). The difference in vertical height between the uppermost points of the condyle (*c*) and the coronoid process (*d*) is therefore unusually small. On the other hand, the lower margin of the bone is undulating, so that it presents a hollow on each side, as well as one near the middle line in front. The two halves of the bone are definitely inclined to one another and this convergence is faintly marked in the two rows of teeth behind the canines. The latter teeth do not project markedly above the level of those adjacent to them. The incisor teeth are remarkably curved in their long axes, with a convexity

PRECURSORS OF PALAEOLITHIC MAN

in front. The prominences called 'genial tubercles' behind the chin are replaced by a shallow pit or fossa.

In one sense the reception accorded by palaeontologists to the fossil jaw of Mauer differs remarkably from most of the comparable instances. That difference consists in the comparative absence of controversy excited by its discovery. This must not be ascribed to any lack of ardour on the part of archaeologists. More probable is it that with the lapse of time, the acceptance of an evolutionary interpretation of the origin of man has gained a wider circle of adherents, so that the claims of even so sensational a specimen as this, are sifted and investigated with a judicial calm much more appropriate and certainly more dignified than the fierce outbursts occasioned by some of the earlier discoveries.

It remains to institute brief anatomical comparisons between the Mauer jaw and those of the highest apes on the one hand, and of the most primitive of human beings on the other.

(a) Of the three larger anthropoid apes available for comparison, it is hard to say which presents the closest similarity. The Gibbons do not appear to approach so nearly as these larger forms. Among the latter, no small range of individual variations occurs. My own comparisons shew that of the material at my disposal the mandible of an Orang-utan comes nearest to the Mauer jaw. But other mandibles of the same kind of ape (Orang-utan) are very different. The chief difficulty in assigning the possessor of the Mauer jaw to a pithecoid stock has been mentioned already. It consists in the inadequate size of the teeth. In addition to this, other evidence comes from the results of an examination of the grinding surfaces (crowns) of the molar teeth. These resemble teeth of the more primitive human types rather than those of apes. Finally the convergence of the two rows when traced towards the canine or eye-tooth of each side, points in the same direction.

(b) If the apes be thus rejected, the next question is, Would the Mauer jaw be appropriate to such a cranium as that of Pithecanthropus? I believe an affirmative answer is justifiable. It is true that an excellent authority (Keith[7]) hesitates on the ground that the mandible seems too massive for the skull, though the same writer recognises that, in regard to the teeth, the comparison is apt. This is a difficult point. For instance the *H. moust. hauseri* (cf. Chapter II) has a mandible which is far 'lower' than the capacity of the brain-case would lead one to expect. Therefore it seems that the degree of correlation between mandible and capacity is small, and to predict the size of the brain from evidence given by the jaw is not always safe. It is to be remembered that special stress was laid by Professor Dubois

1] PRECURSORS OF PALAEOLITHIC MAN 15

(cf. p. 4) on the fact that the teeth of Pithecanthropus when compared with the skull-cap are inadequately small, if judged by the ape-standard of proportion. The characters of the teeth, in so far as upper and lower molars can be compared, present no obstacle to such an association, and in fact provide some additional evidence in its favour. The crucial point seems therefore to be the massiveness of the jaw. With regard to this, the following remarks may be made. First, that the skull-cap of Pithecanthropus is on all sides admitted to shew provision for powerful jaw-muscles. And further, in respect of actual measurements, the comparison of the transverse width of the Javanese skull-cap with that of the Mauer jaw is instructive. For the skull-cap measures 130 mm. in extreme width, the jaw 130 mm. The association of the two does not, in my opinion, make an extravagant demand on the variability in size of either part. A curious comparison may be instituted between the Mauer jaw and the corresponding bone as represented by Professor Manouvrier (cf. Dubois[8], 1896) in an attempted reconstruction of the whole skull of Pithecanthropus. Professor Manouvrier's forecast of the jaw differs from the Mauer specimen chiefly in regard to the size of the teeth, and the stoutness of the ascending ramus. The teeth are larger and the ascending ramus is more slender in the reconstruction than in the Mauer specimen.

(c) Passing from the consideration of Pithecanthropus to that of human beings, the general results of the comparisons that can be made will shew that the gap separating the jaw of Mauer from all modern human representatives is filled by human jaws of great prehistoric antiquity.

The progress of an evolutionary development is accordingly well-illustrated by these specimens. And although *Homo heidelbergensis* is seen to be separated from his modern successors by great differences in form as well as a vast lapse of time, still the intervening period does provide intermediate forms to bridge the gulf. Not the least interesting of many reflections conjured up by the Mauer jaw, is that this extraordinary form should be met with in a latitude so far north of that corresponding to the Javanese discoveries. This difference, together with that of longitude, suggests an immense range of distribution of these ancestral types. Some of their successors are considered in the next chapter.

CHAPTER II

PALAEOLITHIC MAN

THE fossil remains described in the preceding chapter possess good claims to that most interesting position, viz. an intermediate one between Mankind and the more highly-developed of the Apes.

From such remarkable claimants we turn to consider fossil bones of undoubted human nature. Of such examples some have been regarded as differing from all other human types to such an extent as to justify their segregation in a distinct species or even genus. Yet even were such separation fully justified, they are still indubitably human.

In the early phases of the study of prehistoric archaeology, the distinction of a 'stone age' from those of metals was soon realised. Credit is due to the present Lord Avebury[9] for the subdivision of that period into the earlier and later parts known as the Palaeolithic and Neolithic stages. At first, those subdivisions possessed no connotation of anatomical or ethnical significance. But as research progressed, the existence of a representative human type specially characteristic of the palaeolithic period passed from the stage of surmise to that of certainty. Yet, although characteristic, this type is not the only one recognisable in those early days.

In the following pages, some account is given of the most recent discoveries of human remains to which Palaeolithic antiquity can undoubtedly be assigned. The very numerous works relating to prehistoric man are full of discussions of such specimens as those found in the Neanderthal, at Spy, Engis, Malarnaud, La Naulette or Denise.

That some of these examples are of great antiquity is inferred from the circumstances under which they were discovered. The evidence relates either to their association with extinct animals such as the Mammoth, or again the bones may have been found at great depths from the surface, in strata judged to have been undisturbed since the remains were deposited. One of the earliest discoveries was that of the Engis skull; the differences separating this skull from those of modern Europeans are so extraordinarily slight that doubt has been expressed as to the antiquity assigned to the specimen, and indeed this doubt has not been finally dispelled. The bones from Denise (now rehabilitated in respect of their antiquity by Professor Boule) present similar features. But on the other hand the jaws found at La Naulette and Malarnaud suggest the former existence of a lowlier and more bestial form of humanity. Support is provided by the famous skull of the Neanderthal, but in regard to the latter, conclusive evidence (as distinct from presumption) is unfortunately lacking. Further confirmation is given by the Forbes Quarry skull from Gibraltar, but although its resemblance to that of the Neanderthal was clearly noted by Dr Busk and Sir William Turner[10] as long ago as 1864, the specimen was long neglected. In this case, as in that of the Neanderthal, corroborative evidence as to the geological or archaeological horizon is lamentably defective. After a lapse of some twenty years, the discoveries of human skeletons at Spy in Belgium, undoubtedly associated as they were with remains of Mammoth, threw a flood of light on the subject, and enormously enhanced the significance of the earlier discoveries. The former existence in Europe of a human type, different from all other known inhabitants of that continent, and presenting no small resemblance to the lowliest modern representatives of mankind, may be said to have been finally established by the results of the excavations at Spy. Moreover the differences thus recognised are such as to lend strong support to the evolutionary view as to the origin of the more recent human stocks from an ancestral series including representatives of a simian phase. Yet the co-existence of a higher type represented by the Engis skull must not be overlooked, nor indeed has this been the case. The significance of so remarkable a phenomenon is more fully discussed in the sequel; but no detailed account of the earlier discoveries need be given. A bibliography is appended and here references (Hœrnes[44], 1908; Schwalbe[55]) will be found to the more important sources of information upon those specimens.

Locality	Date	Literary reference	Synonyms
Taubach	1895	Nehring[11]	
Krapina	1899	Kramberger[12]	
S. Brélade	1910–11	Marett[13]	
La Chapelle aux Saints	1908	Boule[14]	"Corrèze"
Le Moustier	1908	Klaatsch[15]	"Homo mousterensis hauseri"
La Ferrassie	1909	Peyrony[16]	
Pech de l'Aze	1909	Peyrony[16]	
Forbes Quarry	1848–1909	Sollas[17] Sera[18]	"Gibraltar"
Andalusia	1910	Verner[19]	
Grotte des Enfants	1902–06	Verneau[20]	"Grimaldi"
Baradero	1887	(S. Roth) Lehmann-Nitsche (1907)[21]	
Monte Hermoso	?	Lehmann-Nitsche (1909)[22]	"Homo neogaeus"
Combe Capelle	1909	Klaatsch[23]	"Homo aurignacensis hauseri"
Galley Hill	1895	Newton[24]	"Homo fossilis"

In the present instance, an attempt will be made to provide some account of the most recent advances gained through the results of excavations carried out in late years. And herein, prominence will be given in the first place to such human remains as are assignable to the lowlier human type represented previously by the Spy skeletons. Following upon these, come examples possessing other characters and therefore not referable to the same type.

The discoveries are commonly designated by the name of the locality in which they were made. Those selected for particular mention are enumerated in the list on p. 20.

Taubach in Saxe-Weimar.

Certain specimens discovered at Taubach and first described in 1895 possess an importance second only to that of the Mauer jaw and of the Javan bones found by Professor Dubois. Indeed there would be justification for associating the three localities in the present series of descriptions. But upon consideration, it was decided to bring the Taubach finds into the present place and group. It may be added that they are assigned to an epoch not very different from that represented by the Mauer strata whence the mandible was obtained.

The actual material consists only of two human teeth of the molar series. One is the first lower 'milk' molar of the left side. This tooth exceeds most corresponding modern examples in its dimensions. In a large collection of modern teeth from Berlin no example provided dimensions so large.

The surface is more worn than is usual in modern milk teeth of this kind. The second tooth (Fig. 3) is the first lower 'permanent' molar of the left side. It bears five cusps. Neither this number of cusps, nor its absolute dimensions, confer distinction upon the tooth. Its chief claim to notice is based upon its relative narrowness from side to side. That

Fig. 3. The grinding surface of the first right lower molar tooth from Taubach. The letters denote several small prominences called cusps.

Fig. 4. The grinding surface of the corresponding tooth (cf. Fig. 3) of a Chimpanzee. (Figs. 3, 4, 5, and 6 are much enlarged.)

narrowness (proportion of transverse to antero-posterior diameter), represented by the ratio 84·6:100, is present in a distinctly unusual and almost simian degree. In this character the Taubach tooth resembles the same tooth of the Chimpanzee (Fig. 4), to which it stands nearer than does the corresponding tooth of the Mauer jaw. The manner in which the worn surface of the tooth slopes downwards and forwards has been claimed as another simian character. In these respects, the Taubach tooth is among the most ape-like of human teeth (whether prehistoric or recent) as yet recorded, and in my opinion there is some difficulty in deciding whether

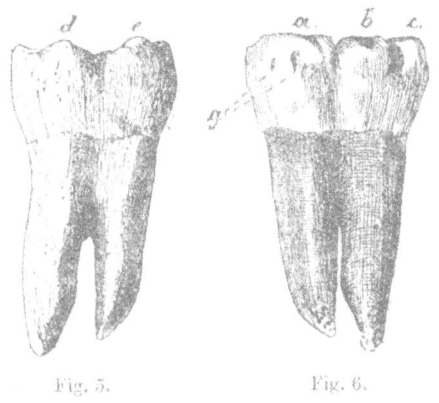

Fig. 5. Inner side of the Taubach tooth.
Fig. 6. Outer side of the same. (From Nehring.)

this is the tooth of a human being or of a pithecoid human precursor. There is a very slight tendency (Figs. 5, 6) to concrescence of the roots, and these are curiously parallel in direction, when viewed from the side. In the latter respect no similarity to the teeth of apes can be recognised.

Krapina in Croatia.

Next in order to the discovery of human teeth at Taubach, the results of excavations in a so-called 'rock-shelter' on the bank of the river Krapinica in Croatia, call for consideration. Immense numbers of bones were obtained, and the remains of a large number of human beings were found to be mingled with those of various animals. Apart from their abundance, the fragmentary character of the human bones is very remarkable. The discovery that one particular stratum in the cave consisted mainly of burnt human bones has suggested that some of the early inhabitants of the Krapina shelter practised cannibalism.

Indeed this view is definitely adopted by Professor Kramberger, and he makes the suggestion that the remains include representatives of those who practised as well as those who suffered from this custom. Both young individuals and those of mature age are represented, but very aged persons have not been recognised.

Turning to the details of the actual bones, the conclusion of outstanding interest is the recognition of further instances of the type of the Neanderthal and of Spy, the latter discovery being separated by a lapse of twenty years and more from that at

Krapina. An attempt has been made to reconstruct one skull, and the result is shewn in Fig. 7, which provides a view of the specimen in profile. Viewed from above, the chief character is the width of the cranial portion, which exceeds very distinctly in this respect the corresponding diameter in the more classic examples from the Neanderthal and

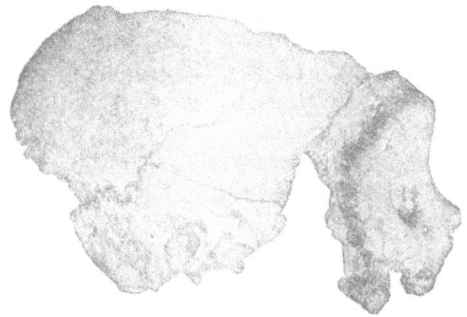

Fig. 7. Profile view of a reconstructed human skull from Krapina. (From Birkner, after Kramberger.)

Spy. It is very important to note that the brain-case is thus shewn to be remarkably capacious, and this is all the more remarkable since the limb-bones do not denote a very great stature or bulk.

Having recently examined the specimens now in the Museum of Palaeontology at Agram in Croatia, I venture to add some notes made on that occasion. The Krapina skull-fragments and the head of a femur are certainly most impressive. It is shewn that early palaeolithic man presents examples of skulls both of brachy-cephalic and dolicho-cephalic proportions. Variations in the form and arrangement of the facial bones also occur.

The form and proportions of the brain-case have been noted already. The profile view (cf. Fig. 7) shews the distinctive features of the brow region. The brow-ridges are very large, but they do not absolutely conform to the conditions presented by the corresponding parts in the skulls of aboriginal Australian or Tasmanian natives. The region of the forehead above the brows is in some instances (but not in all) flattened or retreating, and this feature is indicated even in some small fragments by the oblique direction of the lamina cribrosa of the ethmoid bone.

Two types of upper jaw are distinguishable: no specimen projects forwards so far as might be expected, but the teeth are curiously curved downwards (as in some crania of aboriginal Australians). The facial surface of the jaw is not depressed to form a 'canine fossa.' The nasal bones are flattened.

The mandibles present further remarkable characters. By these again, two types have been rendered capable of distinction. In their massiveness they are unsurpassed save by the mandible from Mauer. In absolute width one specimen actually surpasses the

Mauer jaw, but yet fails to rival that bone in respect of the great width found to characterise the ascending ramus in that example. In the Krapina jaws, the chin is absent or at best feebly developed. In one specimen the body of the jaw is bent at an angle between the canine and first premolar tooth, and is thus reminiscent of the simian jaw. Behind the incisor teeth the conformation is peculiar, again suggestive of the arrangement seen in the Mauer jaw, and differing from that found in more recent human specimens.

The distinction of two types of lower jaw was made in the following manner. The bone was placed on a flat surface. The vertical height of the tooth-bearing part was measured in two regions, (*a*) near the front, (*b*) further back, and close to the second molar tooth (cf. Fig. 2 *f*, *g*). In some of the bones these measurements are nearly equal, but the hinder one is always the less. In the instances in which the two measurements approximate to one another, the proportion is as 100 : 92. In other instances the corresponding proportion differed, the ratio being about 100 : 86 or less. The former type is considered by Professor Kramberger to indicate a special variety (krapinensis) of the Neanderthal or *Homo primigenius* type. The second type is that of the Spy mandible No. 1. Professor Schwalbe[25] (1906) objects to the distinction, urging that the indices (92 and 86) are not sufficiently contrasted. However this may be, it is noteworthy that other bones shew differences. Thus the curvature of the forehead is a variable feature, some skulls having had foreheads much flatter and more retreating than others. The limb bones are also called upon to provide evidence. Some of the arm-bones and thigh-bones are longer and more slender than others.

How far these differences really penetrated and whether the thesis of two types can be fully sustained, does not appear to admit of a final answer. The view here adopted is that, on the whole, the distinction will be confirmed. But nevertheless I am far from supporting in all respects the view of Professor Klaatsch to whose imagination we owe the suggestion of realistic tableaux depicting the murderous conflict of the two tribes at Krapina, the butchery of one act culminating suitably in a scene of cannibalism. Nor am I persuaded that either variety or type found at Krapina can be reasonably identified with that of the Galley Hill skeleton. But of these matters further discussion is reserved for the sequel.

This brief sketch of the cranial characters of the Krapina remains must be supplemented by a note on the teeth. Great numbers were found, and some of them are of enormous dimensions, surpassing

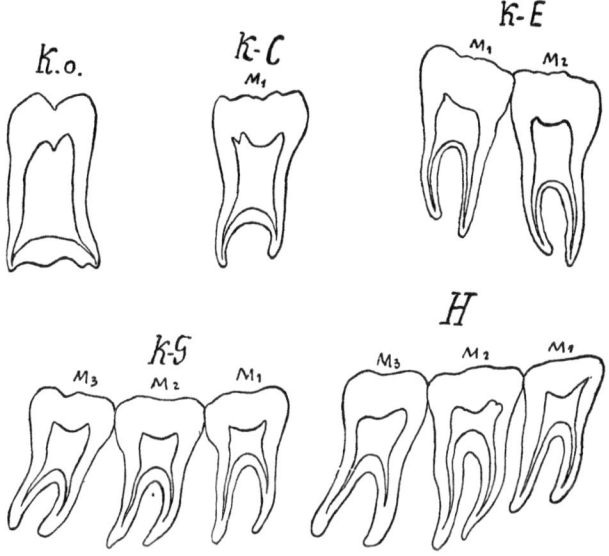

Fig. 8. Tracings (from skiagrams) of various molar teeth. The specimen *K.o.* from Krapina shews the conjoined roots characteristic of teeth found at Krapina, and in Jersey at S. Brélade's Bay. The large pulp-cavity of the Krapina teeth should be noted. *K.o.*, *K.C.*, *K.E.*, *K.G.*, from Krapina; *H.* Mauer. (From Kramberger.)

those of the Mauer jaw. But some of the molar teeth are further distinguished in a very remarkable way, for the roots supporting the crown of the tooth are conjoined or fused: they are not distinct or divergent as is usual. The contrast thus provided by these anomalous teeth is well illustrated in the accompanying figure (8, *Ko*). Now such fusion of roots is not absolutely unknown at the present day; but the third molar or wisdom tooth is most frequently affected. The occurrence is extremely unusual in the other molar teeth of modern men. Yet among the Krapina teeth, such fusion is striking both in its degree and in its frequency. So marked a characteristic has attracted much attention. Professor Kramberger holds the view that it constituted a feature of adaptation peculiar to the Palaeolithic men of Krapina. In opposition to this, Professor Adloff holds that the character is so definite and marked as to enter into the category of distinctive and specific conformations. The discussion of these views was carried on somewhat warmly, but yet to some extent fruitlessly so long as the only known examples were those from Krapina. Dr Laloy supported Professor Kramberger, and on the other side may be ranged the support of Professor Walkhoff. But a recent discovery has very substantially fortified the view adopted by Professor Adloff and his supporters. For in a cave near S. Brélade's Bay in Jersey, the explorations of Messrs Nicolle, Sinel and Marett (1910—1911) have brought to light Palaeolithic human teeth of very similar form. They are said indeed by Dr Keith to be precisely comparable to those from Krapina. The conjoined roots of such teeth should be regarded therefore as more than a peculiarity of the Palaeolithic men of Croatia, and rather as a very definite means of assigning to a particular Palaeolithic epoch any other instances of a similar nature. Space will not admit of more than a simple record of two other features of the Krapina teeth. They are (a) the curvature of the canine teeth and (b) the remarkable size and extent of the 'pulp-cavity' (cf. Fig. 8, *Ko*) of the molar teeth. In entering upon so protracted a discussion of this part of the evidence, the excuse is proffered that, as may be noted in the instances at Trinil and Taubach, teeth are remarkably well-fitted for preservation in the fossil state, since they may be preserved in circumstances leading to the complete destruction of other parts of the skeleton.

The limb bones of the Krapina skeletons are chiefly remarkable for the variety they present. Some are short and stout, of almost pygmy proportions: others are long and slender, inappropriate in these respects to the massive skull fragments which predominate. The distinction of two human types upon evidence furnished by the limb bones has already been mentioned.

S. Brélade's Bay, Jersey.

A cave in this locality has been explored during the last two years (1910, 1911). Human remains are represented by the teeth already mentioned on account of their resemblance to those found at Krapina. The resemblance depends primarily upon the curious fusion of the roots in the molar teeth. Moreover, the circumference of the combined and thickened roots is so great as to confer a most remarkable 'columnar' appearance on the affected teeth (cf. fig. 8, $K.o.$). The teeth from Krapina and Jersey while thus associated must be contrasted with some specimens which they resemble in other respects. The corresponding teeth in the Mauer jaw have been described as similar to those from Krapina, but I cannot confirm this from Dr Schoetensack's illustrations, of which fig. 8 (H) is a fair representation. The teeth of the Forbes Quarry and Le Moustier specimens do not conform to the precise requirements of the test. The Spy teeth are said to have three distinct roots save in two cases, where the numbers are four and two respectively. The test of combined molar roots therefore provides a means of subdividing a group of examples otherwise similar, rather than a mark of recognition applicable to all alike.

The S. Brélade teeth also resemble those from Krapina in the proportions of their crowns and the unusually large size of the pulp-cavity. The latter character may prove more important than the fusion

PALAEOLITHIC MAN

of the roots. But the evidence of their surroundings assigns the teeth from Jersey to an epoch less ancient than that of the Krapina men.

La Chapelle-aux-Saints (Corrèze).

The human skeleton from La Chapelle-aux-Saints

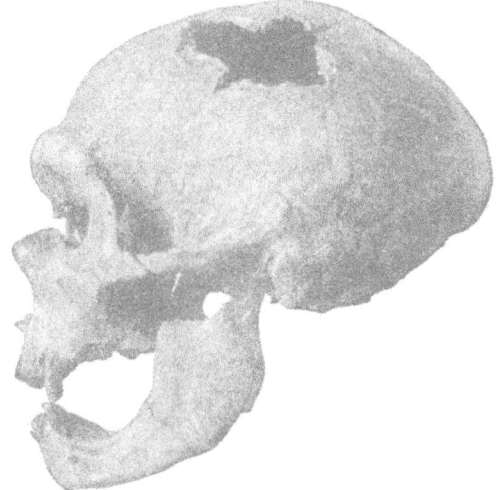

Fig. 9. Profile view of the skull from La Chapelle-aux-Saints (Corrèze). (From Birkner, after Boule.)

holds a very distinguished position among its congeners. In the first place, the discovery was not haphazard, but made by two very competent observers during their excavations. Again; the remains comprise not only the nearly intact brain-case, but much of the facial part of the skull, together with the lower jaw and many bones of the trunk and limbs. The individual was a male of mature age, but not senile (Manouvrier). For these reasons, the value of this skeleton in evidence is singularly great.

Speaking generally, the specimen is found to resemble very closely the Neanderthal skeleton in practically every structure and feature common to the two individuals. This correspondence is confirmatory therefore of the view which assigns great antiquity to the Neanderthal man, and in addition to this, further support is given to the recognition of these examples (together with those from Spy and Krapina) as representatives of a widely distributed type. It is increasingly difficult to claim them as individual variations which have been preserved fortuitously.

Beyond these inferences, the skeleton from La Chapelle adds very greatly to the sum total of our knowledge of the structural details of these skeletons. For here the facial bones are well preserved. Before proceeding to their consideration reference should be made to the side view of the skull (Fig. 9), as well as to the tracings of the brain-case brought into comparison with those provided by the Neanderthal and Spy crania. In the case of one illustration of those tracings (Fig. 10) it must be remarked

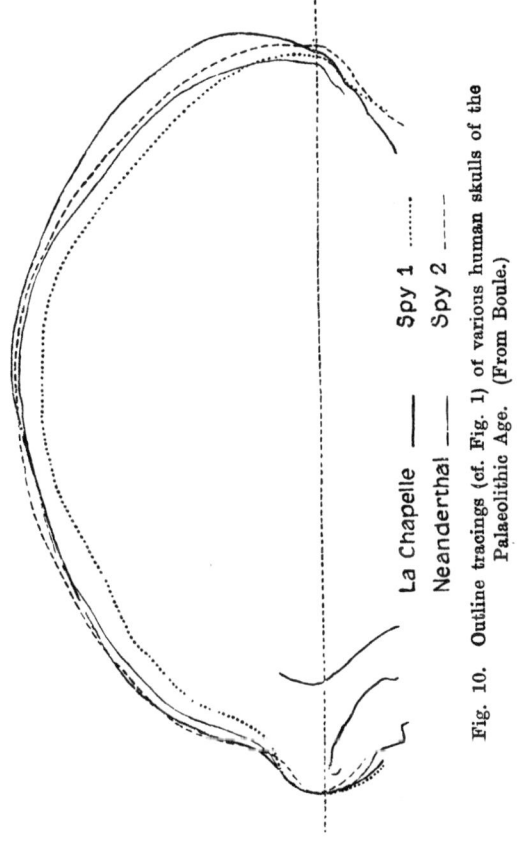

Fig. 10. Outline tracings (cf. Fig. 1) of various human skulls of the Palaeolithic Age. (From Boule.)

that objection is taken by Professor Klaatsch to the base-line selected, though in this particular instance, that objection has less weight than in others.

Turning to the facial parts of the skull, the brows will be seen to overhang the face less than in many crania of aboriginal Australians. Prognathism, *i.e.*

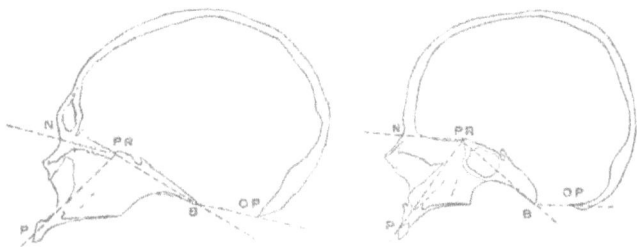

Fig. 11. Contours of two skulls, *A* of a New Guinea man; *B* of an European woman. The angle *B.PR.P* measures the degree of prognathism, and in this respect, the two specimens are strongly contrasted. (From specimens in the Cambridge Museum.)

projection of the jaws (Fig. 11), though distinct, is less pronounced than might be expected. Hereby the reconstruction of the facial parts of the Neanderthal skull, as prepared by Professor Klaatsch, is shewn to be much exaggerated. The skeleton of the nose reveals some simian traits, and on either side, the canine fossa (below the eye) is shallow or non-existent. A good deal of stress has been laid on

PALAEOLITHIC MAN

this character, perhaps more than is justifiable. Yet it is quite uncommon in this degree among modern European crania, though alleged by Giuffrida Ruggeri to characterise certain skulls from the Far East. The reconstructed skull contains teeth which are large and in the incisor region (*i.e.* in front) are much curved downwards in the direction of their length. But this, though probably correct, is yet a matter of inference, for only a couple of teeth (the second premolars of the left side) were found *in situ*. And so far no detailed description of these teeth has appeared. The mandible is of extraordinary dimensions; very widely separated 'ascending rami' converge to the massive body of the jaw. The sigmoid notch is almost as shallow as in the Mauer jaw. The chin is retreating or absent.

Such are the more easily recognisable features of the skull. It will be understood that many more details remain for discussion. But within the allotted space, two only can be dealt with. The capacity of the brain-case is surprisingly large, for it is estimated at 1600 cubic centimetres: from this figure (which will be the subject of further discussion in the sequel) it appears that the man of La Chapelle was amply provided with cerebral material for all ordinary needs as judged even by modern standards. In the second place, MM. Boule and Anthony, not content with a mere estimate of capacity, have published an elaborate account of the form of the brain as revealed by a cast of the interior of the brain-case. As the main result of their investigations, they are enabled to record a list of characters indicative of a comparatively lowly status as regards the form of the brain, although in actual size it leaves little to be desired.

The principal points of interest in the remainder of the skeleton refer in the first instance to the estimate of stature and the evidence provided as to the natural pose and attitude of the individual. Using Professor Pearson's table, I estimate the stature as being from 1600 to 1620 mm. (5ft. 3in. or 5ft. 4in.), a result almost identical with the estimate given for the Neanderthal man. In both, the limb bones are relatively thick and massive, and by the curvature of the thigh-bones and of the upper parts of the shin-bones, a suggestion is given of the peculiar gait described by Professor Manouvrier as 'la marche en flexion'; the distinctive feature consists in an incompleteness of the straightening of the knee-joint as the limb is swung forwards between successive steps.

The bones of the foot are not lacking in interest, and, in particular, that called astragalus is provided with an unusually extensive joint-surface on its outer aspect. In this respect it becomes liable to comparison with the corresponding bone in the feet of climbing animals, whether simian or other.

II] PALAEOLITHIC MAN 39

That these features of the bone in question are not peculiar to the skeleton from La Chapelle, is shewn by their occurrence in bones of corresponding antiquity from La Quina (Martin, 1911) and (it is also said) from La Ferrassie (Boule, L'Anthropologie, Mai-Juin, 1911).

Homo mousterensis hauseri (*Dordogne*)

This skeleton was discovered in the lower rock-shelter of Le Moustier (Dordogne, France) in the course of excavations carried out by Professor Hauser (of Swiss nationality) during the year 1908. The final removal of the bones was conducted in the presence of a number of German archaeologists expressly invited to attend. The omission to inform or invite any French archaeologists, and the immediate removal of the bones to Breslau, are regrettable incidents which cast a shadow quite unnecessarily on an event of great archaeological interest. By a curious coincidence this took place a few days after the discovery of the human skeleton of La Chapelle (*v. supra*). The two finds are very fortunately complementary to each other in several respects, for the Dordogne skeleton is that of a youth, whereas the individual of La Chapelle was fully mature. In their main characters, the two skeletons are very similar, so that in the present account it will be necessary only to mention the more important features revealed by the study of the Dordogne

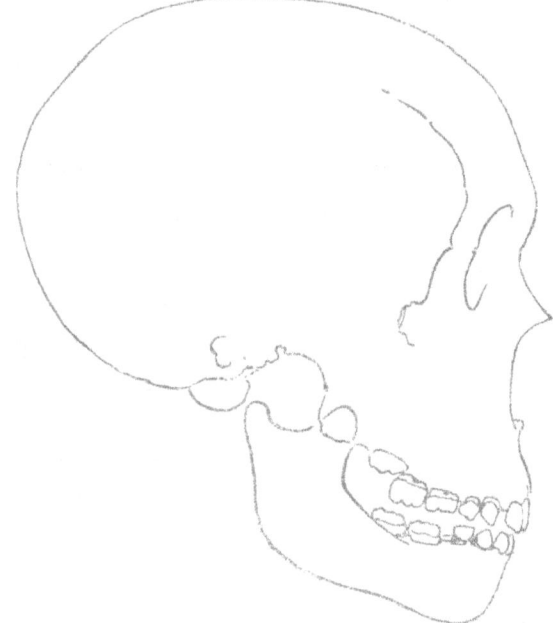

Fig. 12. Outline tracing of a cast of the Moustier skull (Dordogne). (From a specimen in the Cambridge Museum.)

specimen. Outline drawings of the two skulls are compared with the corresponding contour of the Neanderthal calvaria by Klaatsch.

In the Dordogne youth the bones were far more fragile than in the older man from La Chapelle. Nevertheless, photographs taken while the bones were still *in situ* but uncovered, provide a means of realising many features of interest. Moreover although the face in particular was greatly damaged, yet the teeth are perfectly preserved, and were replaced in the reconstructed skull of which a representation

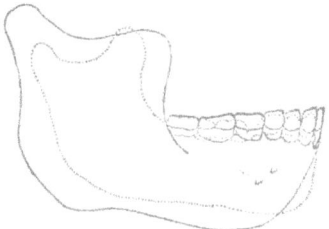

Fig. 13. Tracings from casts (in the Cambridge Museum) of the jaw-bone from Mauer and of that of the Moustier skeleton. The Mauer jaw is indicated by the continuous line.

is shewn in Fig. 12. This reconstruction cannot however be described as a happy result of the great labour bestowed upon it. In particular it is almost certain that the skull is now more prognathous than in its natural state. Apart from such drawbacks the value of the specimen is very great, and this is especially the case in regard to the teeth and the lower jaw. The former are remarkably large, and they agree herein with the teeth from Krapina (though their roots are distinct and not conjoined as

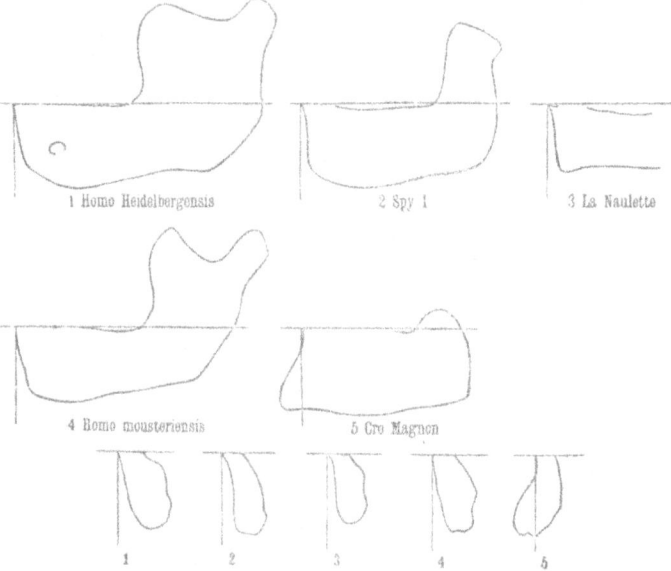

Fig. 14. Outline tracings of jaw-bones. In the lower row, sections are represented as made vertically in the median plane through the chin, which is either receding or prominent. In this series, the numbers refer to those given in the upper set. (From Frizzi.)

in the Krapina examples). In respect of size, the teeth of the Dordogne individual surpass those of the Mauer jaw, but the first lower molar has proportions

similar to the corresponding tooth of that specimen.
But, large as they are, the lower teeth are implanted
in a mandible falling far short of the Mauer jaw in
respect of size and weight (Fig. 13). In fact one of
the great characteristics of the Dordogne skeleton
is the inadequacy of the mandible when compared
to the remainder of the skull, even though allowance
is made for the youth of the individual. Were it

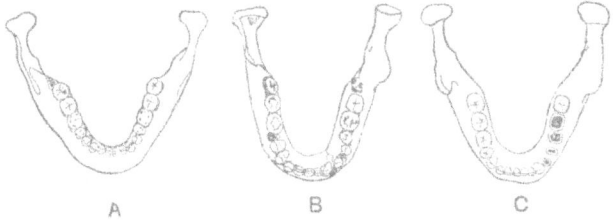

Fig. 15. Outline tracings of jaw-bones viewed from above. *A* an
ancient Briton (cf. Fig. 2, *B*). *B* Moustier. *C* Mauer. (*B* and
C are from casts in the Cambridge Museum.)

not that the facts are beyond dispute, it is difficult to
imagine that such a mandible could be associated
with so large and capacious a cranium. And yet the
jaw is not devoid of points in which it resembles
the Mauer bone, in spite of its much smaller bulk.
Thus the chin is defective, the lower border undulating,
and the ascending branch is wide in proportion to its
height. A good idea of these features is provided by
the illustration of the side-view (cf. Fig. 14) given by

Professor Frizzi. Seen from above, the contour is in close agreement with that of several well-known examples, such as the jaws from Spy (cf. Fig. 15) and Krapina.

The limb bones agree in general appearance with those of the skeletons of the Neanderthal and La Chapelle. Though absolutely smaller than in those examples, they are yet similar in regard to their stoutness. The femur is short and curved, and the articular ends are disproportionately large as judged by modern standards. The tibia is prismatic, resembling herein the corresponding bone in the Spy skeleton. It is not flattened or sabre-like, as in certain other prehistoric skeletons.

Another point of interest derived from the study of the limb bones is the stature they indicate. Having regard to all the bones available, a mean value of about 1500 mm. (about 4 ft. 11 in.) is thus inferred. Yet the youth was certainly 16 years of age and might have been as much as 19 years. The comparison of stature with that of the other examples described is given in a later chapter. At present, it is important to remark that in view of this determination (of 4 ft. 11 in.) and even when allowance is made for further growth in stature the large size of the skull must be regarded as very extraordinary indeed. A similar remark applies to the estimate of the capacity of the brain-case. A moderate estimate gives 1600 c.c. as the capacity of the brain-case (practically identical with that of the La Chapelle skull). In modern Europeans of about 5ft. 6in., this high figure would not cause surprise. In a modern European of the same stature as the Dordogne man (4 ft. 11 in.), so capacious a brain-case would be regarded if not as a pathological anomaly, yet certainly as the extreme upper limit of normal variation. Without insisting further on this paradoxical result (which is partly due to defective observations), it will suffice to remark that early Palaeolithic man was furnished with a very adequate quantity of brain-material, whatever its quality may have been. In regard to the amount, no symptom or sign of an inferior evolutionary status can be detected.

La Ferrassie (Dordogne, France).

This discovery was made in a rock-shelter during its excavation in the autumn of 1909 by M. Peyrony. A human skeleton was found in the floor of the grotto, and below strata characterised by Mousterian implements. The bones were excessively fragile, and though the greatest care was taken in their removal, the skull on arrival at Paris was in a condition described by Professor Boule (L'Anthropologie, 1911, p. 118) as 'très brisée.' No detailed account has yet appeared, though even in its fragmentary condition, the specimen is sure to provide valuable information. From the photographs taken while the skeleton lay *in situ* after its exposure, it is difficult to arrive at a definite conclusion as to its characters. But in regard to these, some resemblance at least (in the jaws) to the Neanderthal type can be detected.

M. Peyrony found also in the same year and in the same region (at Le Pech de l'Aze) the cranium of a child, assignable to the same epoch as the skeleton of La Ferrassie. But so far no further details have been published.

Forbes Quarry (Gibraltar).

The human skull thus designated was found in the year 1848. It was, so to speak, rediscovered by Messrs Busk and Falconer. The former authority described the specimen in 1864, but this description is only known from an abstract in the Reports of the British Association. Broca published an account of the osteological characters a few years later. After 1882, the skull again fell into obscurity for some twenty years: thereafter it attracted the attention of Dr Macnamara, Professor Schwalbe, and above all of Professor Sollas, who published the first detailed and critical account in 1907. This has stimulated yet other researches, particularly those of Professor Sera (of Florence) in 1909, and the literature thus growing up bids fair to rival that of the

Neanderthal skeleton. A most important feature of the specimen consists in the fact that the bones of the face have remained intact and in connection with the skull. But the mandible is wanting, and the molar teeth of the upper set are absent.

As may be gathered from the tracing published by Dr Sera (cf. Fig. 16) the upper part of the brain-case is imperfect. Nevertheless the contour has been restored, and the Neanderthal-like features of distinct brow-ridges, followed by a low flattened cranial curve, are recognisable at once. The facial profile is almost complete, and in this respect the Forbes Quarry skull stood alone until the discovery of the specimen from La Chapelle. Since that incident, this distinction is not absolute, but the Forbes Quarry skull is still unique amidst the other fossils in respect of the bones forming what is called the cranial base. In no other specimen hitherto found, are these bones so complete, or so well preserved in their natural position.

The Forbes Quarry skull is clearly of Neanderthaloid type as regards the formation of the brain-case; in respect of the face it resembles in general the skull from La Chapelle. But in respect of the estimated capacity of the brain-case (estimated at 1100 c.c.), the Forbes Quarry skull falls far short of both those other examples. Moreover the cranial base assigns to it an extremely lowly position. The individual

48 PREHISTORIC MAN [CH.

Fig. 16. Outline tracing and sectional view of the Gibraltar (Forbes Quarry) skull. The various angles are used for comparative purposes. (From Sera.)

PALAEOLITHIC MAN

is supposed by some to have been of the female sex, but there is no great certainty about this surmise. The enormous size of the eye-cavities and of the opening of the nose confer a very peculiar appearance upon the face, and are best seen in the full-face view. Some other features of the skull will be considered in the concluding chapter, when its relation to skulls of the Neanderthal type will be discussed in detail.

Andalusia, Spain.

In 1910, Colonel Willoughby Verner discovered several fragments of a human skeleton in a cave in the Serranía de Ronda. These fragments have been presented to the Hunterian Museum. They seem to be absolutely mineralised. Though imperfect, they indicate that their possessor was adult and of pygmy stature. The thigh-bone in particular is of interest, for an upper fragment presents a curious conformation of the rounded prominence called the greater trochanter. In this feature, and in regard to the small size of the head of the bone, the femur is found to differ from most other ancient fossil thigh-bones, and from those of modern human beings, with the exception of some pygmy types, viz. the dwarf-like cave-dwellers of Aurignac (compared by Pruner-Bey in 1868 to the Bushmen), the aborigines of the Andaman islands, and the aboriginal Bushmen of South Africa. A full description of the bones has not been published, but will probably appear very shortly.

Grimaldi (Mentone Caves).

Among the numerous human skeletons yielded by the caves of Mentone, two were discovered at a great depth in a cave known as the 'Grotte des Enfants.' The excavations were set on foot by the Prince of Monaco, and these particular skeletons have been designated the 'Grimaldi' remains.

Their chief interest (apart from the evidence as to a definite interment having taken place) consists in the alleged presence of 'negroid' characters. The skeletons are those of a young man (cf. Fig. 17), and an aged woman. The late Professor Gaudry examined the jaw of the male skeleton. He noted the large dimensions of the teeth, the prognathism, the feeble development of the chin, and upon such grounds pointed out the similarity of this jaw to those of aboriginal natives of Australia. Some years later Dr Verneau, in describing the same remains, based a claim to (African) negroid affinity on those characters, adding thereto evidence drawn from a study of the limb bones. In both male and female alike, the lower limbs are long and slender, while the forearm and shin-bones are relatively long when compared respectively with the arm and the thigh-bones.

From a review of the evidence it seems that the term 'negroid' is scarcely justified, and there is no

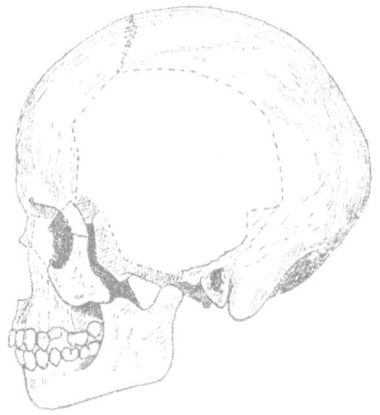

Fig. 17. Profile view of young male skull of the type designated that of 'Grimaldi,' and alleged to present 'negroid' features. *Locality*. Deeper strata in the Grotte des Enfants, Mentone. (From Birkner, after Verneau, modified.)

doubt that the Grimaldi skeletons could be matched without difficulty by skeletons of even recent date. Herein they are strongly contrasted with skeletons of the Neanderthal group. And although modern

Europeans undoubtedly may possess any of the osteological characters claimed as 'negroid' by Dr Verneau, nevertheless the African negro races possess those characters more frequently and more markedly. Caution in accepting the designation 'negroid' is therefore based upon reluctance to allow positive evidence from two or three characters to outweigh numerous negative indications; and besides this consideration, it will be admitted that two specimens provide but a feeble basis for supporting the superstructure thus laid on their characters. Lastly Dr Verneau has been at some pains to shew that skulls of the 'Grimaldi-negroid' type persist in modern times. Yet the possessors of many and probably most such modern crania were white men and not negroes.

Enough has however been related to shew how widely the skeletons from the 'Grotte des Enfants' differ from the Palaeolithic remains associated as the Neanderthal type.

South America. With the exception of Pithecanthropus, all the discoveries mentioned in the foregoing paragraphs were made in Europe. From other parts of the world, actual human remains referable to earlier geological epochs are scanty save in South America. The discoveries made in this part of the New World have been described at great length. In many instances, claims to extraordinary

PALAEOLITHIC MAN

antiquity have been made on their behalf. It is necessary therefore to examine the credentials of such specimens. Upon an examination of the evidence, I have come to the conclusion that two instances only deserve serious attention and criticism.

Baradero.

Fragmentary remains of a human skeleton: the mandible is the best preserved portion; unfortunately the front part has been broken off so that no conclusion can be formed as to the characters of the chin. Otherwise in regard to its proportions, some resemblance is found with the mandible of the Spy skull (No. 1). More important and definite is the direction of the grinding surfaces of the molar teeth. In the lower jaw, this surface is said to look forwards. The interest of this observation consists in the fact that the tooth from Taubach presents the same feature, which is unusual.

Beyond these, the skeleton from the löss of Baradero presents no distinctive features save the remarkable length of the upper limbs.

Monte Hermoso.

From this region two bones were obtained at different dates. These are an atlas vertebra (the vertebra next to the skull) and a thigh-bone. The latter is of less than pygmy dimensions. Both are from fully adult skeletons.

An attempt has been made to reconstruct an individual (the Tetraprothomo of Ameghino) to which the two bones should be referred. It will be noticed that the circumstances bear some, although a very faint, analogy to those in which the remains of Pithecanthropus were found. The results are however extraordinarily different. Professor Branco has ably shewn that in the case of the bones from Monte Hermoso, the association in one and the same skeleton would provide so large a skull in proportion to the rest of the body, that the result becomes not only improbable, but impossible. It is therefore necessary to treat the bones separately. If this is done, there is no reason to regard the thighbone as other than that of a large monkey of one of the varieties known to have inhabited South America in prehistoric as well as in recent times.

The vertebra is more interesting. It is small but thick and strong in a degree out of proportion to its linear dimensions. Professor Lehmann-Nitsche supposes that it may have formed part of a skeleton like that of Pithecanthropus, that is to say that it is not part of a pygmy skeleton. On the other hand, Dr Rivet considers that the Monte Hermoso vertebra could be matched exactly by several specimens in the large collection of exotic human skeletons in the

National Museum, Paris. Be this as it may, there is no doubt that the atlas vertebra in question constitutes the most interesting discovery of its kind made so far in South America. It is important to notice that time after time the attempts made to demonstrate the early origin of Man in the American Continent have resulted in failure, which in some instances has been regrettably ignominious.

Combe Capelle (**H. aurignacensis hauseri**).

Returning to Europe, it is to be noted that in a rock-shelter near Combe-Capelle (Dordogne), the excavations of Dr Hauser led to the discovery in 1909 of an entire human skeleton of the male sex. The interment (for such it was) had taken place in the Aurignacian period. The skeleton presents a very striking appearance. In stature, no important divergence from the Neanderthal type can be noted. But the more vertical forehead, more boldly-curved arc of the brain-case, the diminished brow-ridges, large mastoid processes and distinct canine fossae provide a complete contrast between the Aurignac man and those of the Neanderthal group. Moreover the Aurignac jaw has a slight projection at the chin, where an 'internal process' is now distinct. The brain-case has dolicho-cephalic proportions in a marked degree. The limb bones are straight and slender, and not so much enlarged in the regions of the several joints.

The Aurignac skeleton of Combe Capelle has been associated with several others by Professor Klaatsch. By some authorities they are considered as transitional forms bridging the gap between the early Palaeolithic types and those of the existing Hominidae. But Professor Klaatsch evidently regards them as intruders and invaders of the territory previously occupied by the more lowly Neanderthaloid type.

Galley Hill.

Among the skeletons which have been thus associated with the Aurignac man, are three which have for many years attracted the attention of anthropologists. For this reason, no detailed account of their characters will be given here. Of the three instances referred to, two are the fragmentary skull-caps of the skeletons found at Brüx and at Brünn in Moravia. The latter specimen is generally described as Brünn (91) to distinguish it from Brünn (85), a different and earlier find of less interest.

It will suffice to mention here that both specimens agree in possessing what may be described as a distinctly mitigated form of the characters so strongly developed in the Neanderthal skull and its allies.

PALAEOLITHIC MAN

The Aurignac and Brüx skulls are distinctly longer and narrower than that of Brünn (91). The limb bones are not available for the purposes of evidence.

The third specimen possesses a very much greater interest. It is known as the Galley Hill skeleton from the site of its discovery near Northfleet in Kent. Since it was first described by Mr E. T. Newton (in 1895), much literature has accumulated about the difficult problems presented by the Galley Hill skeleton. By some authors it is regarded as clearly associated with the other examples just mentioned (Brüx, Brünn, and Aurignac). Others reject its claims to high antiquity; of the latter some are courteous, others are scornful, but all are absolutely decided. Having investigated the literature as well as I could, and having seen the cranium, I decided that the claims to great antiquity made on its behalf do really justify its inclusion. But I am quite convinced that the skeleton will give no more than very general indications. Thus the bones are fragile in the extreme. And besides this, the skull is so contorted that measurements made in the usual way must be extraordinarily misleading and the possible error is too great to be successfully allowed for (cf. Fig. 18).

To insist upon these points is the more important since nowadays various indices based on such measurements of the Galley Hill cranium will be found tabulated with data yielded by other skulls,

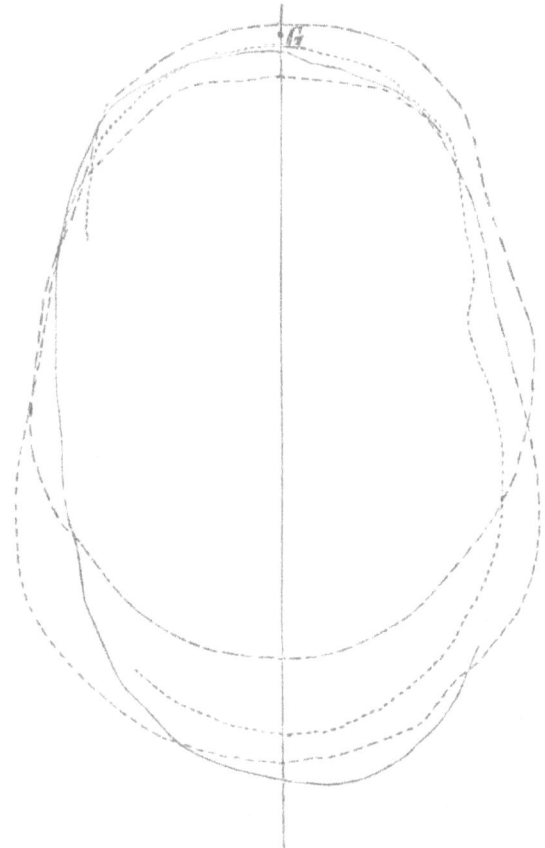

Fig. 18. Outline tracing of the Galley Hill skull, viewed from above. (From Klaatsch.)
—— Galley Hill.   ——— Ancient German.
····· Neanderthal.   ----- Modern South German.

CH. II] PALAEOLITHIC MAN 59

and yet no mark of qualification distinguishes the former figures.

The description of the skeleton may be given in a very few words. In the great majority of its characters, it is not seen to differ from modern human beings (though the stature is small, viz. 1600 mm., 5 ft. 3 in.). And so far as I am able to judge, the characters claimed as distinctive (separating the Galley Hill skull from modern dolichocephalic European skulls) are based upon observations containing a very large possibility of error.

Having regard to such statements, the inference is that the Galley Hill skull does not in fact differ essentially from its modern European counterparts. Similar conclusions have been formed in regard to the other parts of this skeleton. It is important to note that the specimen does not lose its interest on this account.

Summary.

From the foregoing descriptions, it follows that of the most ancient remains considered, at least three divisions can be recognised. In the first place, come the examples described as Pithecanthropus and *Homo heidelbergensis* (Mauer). In the second category come instances as to which no reasonable doubt as to their definitely human characters now exists (save possibly in the case of the Taubach tooth and the Hermoso atlas). Of the members of this second series, two sub-divisions here designated (*A*) and (*B*) can be demonstrated; these with the first examples complete the threefold grouping set out in the table following, with which Table A, p. 85, should be compared.

GROUP I. Early ancestral forms. *Ex. gr. H. heidelbergensis.*
GROUP II.
    Subdivision *A*. *Homo primigenius.* *Ex. gr. La Chapelle.*
    Subdivision *B*. *H. recens*; with varieties $\begin{cases} \text{H. fossilis. } Ex.\,gr. \\ \quad Galley\ Hill. \\ \text{H. sapiens.} \end{cases}$

Taking the first group (Pithecanthropus and *Homo heidelbergensis*) it is to be noticed that close correlation is quite possible. Besides this, evidence exists in each case to the effect that far-distant human ancestors are hereby revealed to their modern representatives. Of their physical characters, distinct indications are given of the possession of a small brain in a flattened brain-case associated with powerful jaws; the lower part of the face being distinguished by the absence of any projection of the chin. The teeth indicate with some degree of probability that their diet was of a mixed nature, resembling in this respect the condition of many modern savage tribes. Beyond this, the evidence is weak and indefinite. It is highly probable that these men were not arboreal:

though whether they habitually assumed the distinctive erect attitude is a point still in doubt. And yet again, while the indications are not clear, it is probable that in stature they were comparable, if not superior, to the average man of to-day.

Passing from this division to the second, a region of much greater certainty is entered. Of the second group, one subdivision (*A*) retains certain characters of the earlier forms. Thus the massive continuous brow-ridge persists, as do also the flattened brain-case with a large mass of jaw-muscle, and a ponderous chinless lower jaw. For the rest, the points of contrast are much more prominent than those of similarity. The brain has increased in size. This increase is very considerable in absolute amount. But relatively also to the size of the possessor, the increase in brain-material is even more striking, for the stature and consequently bulk and weight are less. The thigh-bone offers important points of difference, the earlier long slender form (in *P. erectus*) being now replaced by a shorter, curved, thick substitute. If there has been inheritance here, marked and aberrant variation is also observed.

The second subdivision (*B*) remains for consideration. Here the stature has not appreciably changed. The limb bones are long, slender, and less curved than those of the other associated human beings (*A*), and herein the earliest type is suggested once more.

But the differences occur now in the skull. The brain is as large as in the other subdivision (*A*) and in modern men. The brain-case is becoming elevated: the brow-ridges are undergoing reduction; this process, commencing at their outer ends, expresses to some extent the degree of reduction in the muscles and bone of the lower jaw. The teeth are smaller and the chin becomes more prominent. The distinction from modern types of humanity is often impossible.

In the next chapter some account is given of the circumstances under which the bones were discovered, and of the nature of their surroundings.

CHAPTER III

ALLUVIAL DEPOSITS AND CAVES

THE principal characters of the oldest known human remains having been thus set forth, the circumstances of their surroundings next demand attention. A brief indication of these will be given with the aid of the illustrations provided in the original memoirs in each case, and the order of descriptions followed in the preceding chapter will be observed.

Pithecanthropus. The remains of Pithecanthropus were recovered from an alluvial deposit at Trinil. A section of this is shewn in Fig. 19. An idea may thus be gained of the very considerable amount of superincumbent materials. The associated fauna cannot be compared directly to that of any Western European locality. But in comparison with the modern fauna of Java, the strata in which the Pithecanthropus was found shew a predominance of extinct species, though not of genera. Elephants and hippopotami were present: they point to a close relation between the fauna of Trinil and that of certain Siwalik strata in India, referred to a late

Pliocene age. The difference of opinion upon this point has been mentioned in the preceding chapter:

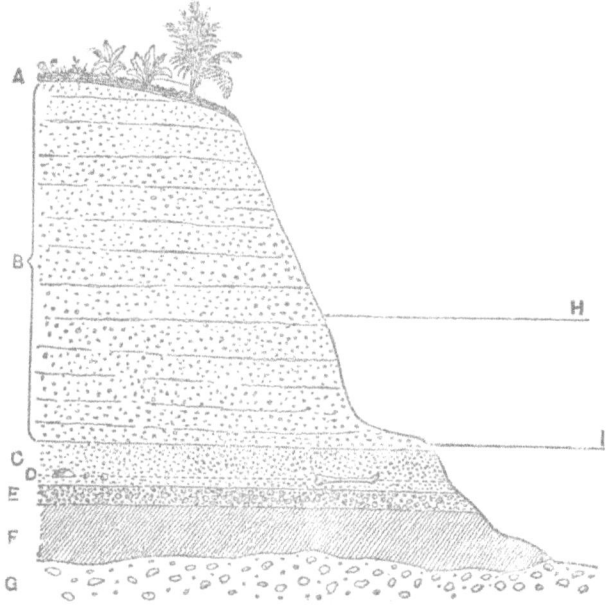

Fig. 19. Section of the strata at Trinil in Java. *A* vegetable soil. *B* Sand-rock. *C* Lapilli-rock. *D* Level at which the bones were found. *E* Conglomerate. *F* Clay. *H* Rainy-season level of river. *I* Dry-season level of river. (From Dubois.)

here it will suffice to repeat that a final conclusion does not appear to have been reached, and that the

III] ALLUVIAL DEPOSITS AND CAVES 65

experts who have examined the strata *in situ* still differ from each other.

Mauer. Impressed by the similarity of the conditions at Mauer to those of the fossiliferous tufa-beds near Taubach and Weimar, Dr Schoetensack had anticipated the possibility of obtaining valuable fossil relics from the former locality. For some twenty years,

Fig. 20. View of the Mauer sand-pit. X (in white) position of jaw-bone when found. (From Birkner, after Schoetensack.)

Dr Schoetensack kept in touch with the workmen of Mauer, and thus when the jawbone was found, he was summoned at once. Even so, the jaw had been removed from its resting-place, and broken in two fragments. Yet there is no doubt as to the exact position in which it was found. Sand and löss (a fine earthy deposit) had accumulated above it to a thickness of seventy feet. The nature of the surroundings may be estimated by reference to the illustration (Fig. 20) reproducing Dr Schoetensack's photograph of the sand-pit. The sands which contained the mandible represent an alluvial deposit, and so far resemble the Trinil beds in Java. The attempt to institute an exact comparison would be unprofitable, but on the whole it would seem that, of the two, the Mauer sands represent the later stage. The fauna associated with the Mauer jaw includes such forms as *Elephas antiquus, Rhinoceros etruscus, Ursus arvernensis, U. deningeri* (an ancestral form of *U. spelaeus*), together with a species of horse intermediate between *Equus stenonis*, and the fossil horse found at Taubach. The cave-lion, bison, and various deer have also been recognised.

The aspect of this collection shews a marked similarity to that of the so-called Forest-bed of Cromer, though at the same time indicating a later age. The Mauer jaw must therefore be assigned to the very earliest part of the Pleistocene epoch. In his original memoir, Dr Schoetensack gave no account of any associated 'industry,' in the form of stone implements. But now (1911) Professor Rutot unhesitatingly (though the reasons are not stated) ascribes to the horizon of the Mauer jaw, that division of the eolithic industries termed by him the "Mafflien." Upon the correctness of such a view judgment may well be reserved for the present.

III] ALLUVIAL DEPOSITS AND CAVES 67

Taubach. The bone-bed (*Knochenschicht*) of Taubach whence the two human teeth were recovered, lies at a depth of some 15 feet (5·2 m.) from the adjacent surface-soil. No fewer than eleven distinct horizons have been recognised in the superincumbent strata. Palaeoliths had often been obtained from the same stratum as that which yielded the human teeth. Dr Weiss referred it to the first, *i.e.* the earlier of two inter-glacial periods judged to have occurred in this region. The associated fauna includes *Elephas antiquus, Rhinoceros merckii, Bison priscus*, with Cervidae and representatives of swine, beaver and a bear. The similarity of this assemblage to that of the Mauer Sands has been noted already.

The hippopotamus however does not seem to have been recorded in either locality. Nevertheless, the general aspect of the mammalian fauna is 'southern' (*faune chaude* of French writers). Upon this conclusion, much depends, for the Palaeolithic implements (claimed as contemporaneous with the extinct 'southern' mammals recorded in the foregoing paragraphs) are said to correspond to the type of Le Moustier. But Mousterian implements are (it is alleged) practically never associated with 'southern' animals, so that in this respect the Taubach bone-bed provides a paradox. Without discussing this paradox at length, it may be stated that the implements just described as 'Mousterian' are not recognised as such by all the experts. Thus Obermaier identifies them with those of Levallois, *i.e.* a late S. Acheul type (cf. Obermaier, 1909). Others declare that the type is not that of Le Moustier, but of Chelles. The latter type of implement is found habitually in association with the southern fauna, and thus the paradox described above may prove to be apparent only and not real. But the unravelling of the different opinions relating to the Taubach finds is among the easier tasks presented to anyone desirous of furnishing a clear statement of the actual state of our knowledge on these matters. The difficulties with which the whole subject bristles may thus be realised.

Krapina. Researches productive of evidence as to the existence of Palaeolithic man in Croatia, were commenced at Krapina so long ago as August, 1899, by Professor Kramberger. A preliminary report was published in December, 1899. Until the year 1904 these researches passed almost unnoticed in this country. The site was not exhausted until 1905. The actual excavations were made in a rock-shelter on the right bank of the Krapinića river, near the village of Krapina. The rock-shelter had been to some extent invaded not long before the archaeological work commenced, and evidence of early human occupation of the site was revealed in the form of

III] ALLUVIAL DEPOSITS AND CAVES 69

dark bands of earth, containing much charcoal. These bands were seen as lines in the lower parts of the exposed section of the cave contents. Fragments

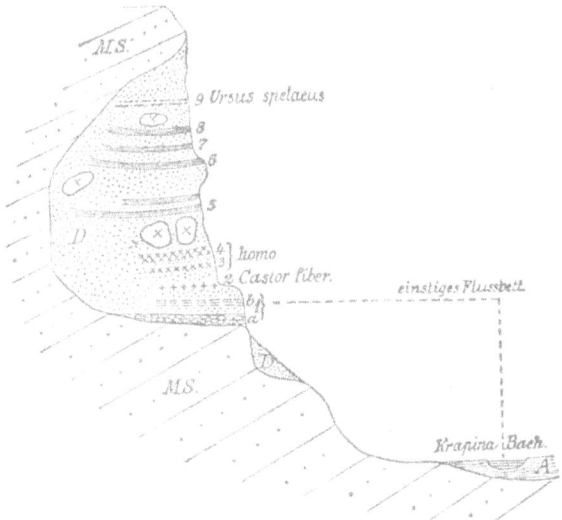

Fig. 21. Section of the Krapina rock-shelter. 3, 4 strata with human remains. 1 b former level of river-bed. (From Birkner, after Kramberger.)

of human and other bones to the number of several thousands were removed. In one season's work six hundred stone implements were found.

A section of the several strata has been published and is reproduced in Fig. 21. Human bones or artefacts were found throughout a wide series of strata, in which no variations of a cultural nature were detected. Throughout the period of human occupation, the Palaeolithic inmates of the cave remained on an unaltered and rather lowly level of culture. This is described by some authorities as Mousterian, by others as Aurignacian; in either case as of an early Palaeolithic aspect.

But when the animal remains are considered, Krapina seems to present the difficulty already encountered in the case of Taubach. For there is no doubt but that the 'southern' fauna is to some extent represented at Krapina. This qualified form of statement is employed because one representative only, viz. *Rhinoceros merckii*, has been discovered, whereas its habitual companions, *Elephas antiquus* and Hippopotamus, have left no traces at Krapina. Other animals associated with the cave-men of Krapina are not so commonly found in the presence of the *Rhinoceros merckii*. Thus the *Ursus spelaeus, U. arctos, Bos primigenius*, and the Arctomys (Marmot) are suggestive of a more northern fauna. But the presence of even a possibly stray *Rhinoceros merckii* is sufficient to confer an aspect of great antiquity on this early Croatian settlement. No evidence of formal interments has come to light, and

III] ALLUVIAL DEPOSITS AND CAVES 71

as regards the cannibalistic habits of the human cave-dwellers, no more than the merest surmise exists.

S. Brélade's Bay, Jersey. In the cave thus designated, old hearths were met with at a depth of twenty-five feet below the surface. Human beings are represented by teeth only. No evidence of interments has been recorded. The implements are of Mousterian type. Associated with the hearths and implements were many fragmentary remains of animals. Up to the present time, the following forms have been identified: *Rhinoceros tichorhinus* (the hairy rhinoceros), the Reindeer, and two varieties of Horse. So far as this evidence goes, the age assigned to the implements is supported, or at least not contra-indicated. It is most improbable that the period represented can be really earlier than the Mousterian, though it might be somewhat later. That the Krapina teeth (which so curiously resemble those of S. Brélade's Bay in respect of the fusion of their roots) should be assigned to the same (Mousterian) epoch is perhaps significant.

La Chapelle-aux-Saints (Corrèze). This is the best example of an interment referable to the early Palaeolithic age (Fig. 22). Two reasons for this statement may be given. In the first place, the skeleton lay in a distinctly excavated depression, beneath which no signs of an earlier settlement are recorded. Secondly, the superincumbent strata can be assigned

72 PREHISTORIC MAN [CH.

to one period only of the archaeological series, viz. that of Le Moustier. Indications of the preceding period (S. Acheul) as well as of the subsequent one (Aurignac) are practically negligible. Moreover the

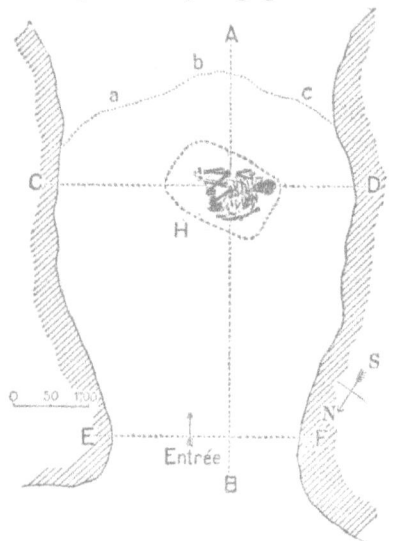

Fig. 22. Plan of the cave at La Chapelle-aux-Saints (Corrèze). [From Boule.]

surroundings had not been disturbed since the interment: this is shewn by the leg-bones of a large bovine animal (Bison or Bos) found in their natural relations just above the head of the human skeleton.

III] ALLUVIAL DEPOSITS AND CAVES 73

The latter lay on the back, the right arm bent, the left extended; both legs were contracted and to the right. In general, this attitude recalls that of the skeletons of La Ferrassie and the Grotte des Enfants (Grimaldi). At Le Moustier too, the skeleton was found in a somewhat similar position.

At La Chapelle-aux-Saints, the associated fauna includes the Reindeer, Horse, a large bovine form (? Bison), *Rhinoceros tichorhinus*, the Ibex, Wolf, Marmot, Badger and Boar.

It would seem that this particular cave had served only as a tomb. For other purposes its vertical extent is too small. The stone artefacts are all perfect tools: no flakes or splinters being found as in habitations. The animal remains are supposed to be relics of a funeral feast (or feasts). But the presence of the Rhinoceros is perhaps antagonistic to such an explanation.

Le Moustier (Dordogne). The skeleton lay on its right side, the right arm bent and supporting the head; the left arm was extended. The stratum upon which the body rested consisted largely of worked flint implements. These are assigned to the later Acheulean and earlier Mousterian epochs.

Two features in contrast with the conditions at La Chapelle are to be noticed. It is doubtful whether the skeleton at Le Moustier had been literally interred. It seems rather to have been placed on what was at the time the floor of the grotto, and then covered partly with earth on which implements were scattered Indications of a definite grave were found at La Chapelle. Again at Le Moustier, other parts of the same grotto had been occupied as habitations of the living. At La Chapelle this seems not to have been the case.

The evidence of the accompanying animal remains also differs in the two cases. At Le Moustier, only small and very fragmentary animal bones with the tooth of an ox were found in the immediate vicinity of the human skeleton. An extended search revealed bones of *Bos primigenius* in the cave. No bones of the Reindeer were found and their absence is specially remarked by Professor Klaatsch, as evidence that the skeleton at Le Moustier is of greater antiquity than the skeleton accompanied by reindeer bones at La Chapelle. In any case, it would seem that no great lapse of time separates the two strata.

La Ferrassie. The skeleton was found in the same attitude as those of La Chapelle and Le Moustier, viz. in the dorsal position, the right arm bent, the left extended, both legs being strongly flexed at the knee and turned to the right side. The bones were covered by some 3·5 m. of *débris*: stone implements were yielded by strata above and below the body respectively. Beneath the skeleton, the implements are of Acheulean type, while above and around it the type of Le Moustier was encountered. Aurignacian implements occurred still nearer the surface.

III] ALLUVIAL DEPOSITS AND CAVES

In regard to the evidence of interment the conditions here resemble those at Le Moustier rather than those of La Chapelle. The human skeleton did not appear to have been deposited in a grave, but simply laid on the ground, covered no doubt by earth upon which flint implements were scattered. But the cave continued to be occupied until at the close of the Aurignacian period a fall of rock sealed up the entrance. It is difficult to realise the conditions of life in such a cave, after the death of a member of the community, unless, as among the cave-dwelling Veddas of Ceylon, the cave were temporarily abandoned (Seligmann, 1911). It is possible that the normal accumulation of animal remains created such an atmosphere as would not be greatly altered by the addition of a human corpse, for Professor Tylor has recorded instances of such interments among certain South American tribes. But it is also conceivable that the enormously important change in custom from inhumation to cremation, may owe an origin to some comparatively simple circumstance of this kind. The animal remains at La Ferrassie include Bison, Stag, and Horse, with a few Reindeer. The general aspect is thus concordant with that at La Chapelle.

Pech de l'Aze. It is impossible to decide whether the child's skull had been buried intentionally or not. The associated fauna is apparently identical with that of La Ferrassie and La Chapelle.

Forbes Quarry (*Gibraltar*). Of the surroundings of the Forbes Quarry skull at the time of its discovery nothing is known. In 1910 the present writer explored Forbes Quarry and a small cave opening into it. But no evidence of the presence of prehistoric man was obtained. Bones of recent mammalia and certain molluscs found during the excavations, throw no light on this subject.

Andalusia. At the time of writing, only the following information is available as to the surroundings of these human cave-bones. They were discovered on or near the floor of a deep fissure leading to a series of labyrinthine passages. The walls of the fissure or cave were decorated with drawings of animals resembling those at Cretas in Aragon. Besides the mineralised bones, other fragments of less antiquated aspect were found. Potsherds were also obtained, but I have no information as to the occurrence of implements.

Grotte des Enfants (*Mentone*). With regard to the two 'negroid' skeletons of this cave, the first important point is the enormous thickness of accumulated *débris* by which the bones were covered. A depth of some twenty-four feet had been reached before the discovery was made (Fig. 23).

The bodies had been definitely interred, large stones being found in position, adjusted so as to protect the heads particularly. The bodies had been

III] ALLUVIAL DEPOSITS AND CAVES 77

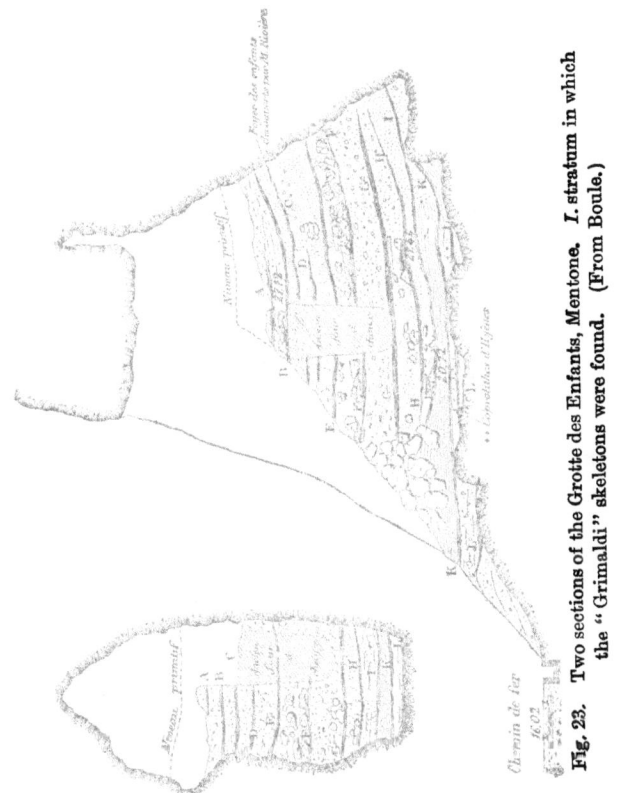

Fig. 23. Two sections of the Grotte des Enfants, Mentone. I. stratum in which the "Grimaldi" skeletons were found. (From Boule.)

placed on the right side. Of the woman, both arms were bent as were the lower limbs. The male skeleton has the right arm flexed, but the left extended (as in the cases of La Chapelle, Le Moustier, and La Ferrassie).

It is practically certain that the skeletons do not belong to an epoch represented, as regards its culture or fauna, by strata lower than that which supported the human remains. This conclusion is very important here. For the evidence of the stone implements accompanying the human bones is fairly definite: it points to the Mousterian age. The animal bones are those of the Reindeer and Cave Hyaena. The presence of the former animal supports the conclusion arrived at on the evidence of the human artefacts. The presence of the Cave Hyaena does not controvert that conclusion.

But an interesting fact remains to be considered. Below the two human skeletons, the animal remains are those of the 'southern' fauna. All the characteristic representatives were found, viz. *Elephas antiquus*, *Rhinoceros merckii*, and Hippopotamus. The Hyaena was also associated with these large animals. It is not clearly stated whether implements of Mousterian type occurred in these, the deepest strata of the cave-floor. Were this so, the contention made in respect of the Taubach implements (cf. *supra*, p. 67) would be remarkably corroborated, as would also the somewhat similar suggestion made in regard to Krapina. For the moment, however, it must suffice to attribute these human remains of negroid aspect to the Mousterian period at Mentone. Inasmuch as the reindeer appears in several strata overlying the remains of the Grimaldi race (for so it has been named by Dr Verneau), it is certainly conceivable that the two individuals are Aurignacian or even later. But this is to enter a wilderness of surmise. Human skeletons were actually found in those more superficial strata and also were associated with the Reindeer. But their cranial features are of a higher type (Cro-Magnon) and contrast very clearly with those of the more deeply buried individuals.

South America. The two discoveries mentioned in the preceding chapter, were made in the so-called Pampas formation of Argentina. This formation has been subdivided by geologists into three successive portions, viz. upper, middle and lower. The distinction is based partly upon evidence derived from the actual characters of deposits which differ according to their level. But the molluscan fauna has also been used as a means of distinction. The whole formation is stated by some to be fluviatile. Other observers speak of it as Löss. This need not necessarily exclude a fluviatile origin, but speaking generally that term now suggests an aerial rather than a subaqueous deposit. The upper subdivision is designated the yellow löss in contrast to the brown löss forming the middle layer. Opinion is much divided as to the exact geological age of the Pampas formation. Ameghino refers it to the Pliocene period, excepting the lower divisions which he regards as upper Miocene. Professor Lehmann-Nitsche assigns Pliocene antiquity to the lowest sub-division only. Dr Steinmann regards the middle and lower subdivisions as equivalents of the 'older' löss of European Pleistocene deposits. The latter determinations are more probably correct than is the first.

Baradero. The Baradero skeleton was obtained from the middle formation or brown löss, in a locality marked by the presence of mollusca corresponding with modern forms, and contrasted with the Tertiary Argentine mollusca. The skeleton was in a 'natural' (*i.e.* not a contracted) position, the head being depressed on the front of the chest. No associated implements or remains of mammalian skeletons are recorded.

Monte Hermoso. The vertebra and femur were found in the lower subdivision of the Pampas formation. We have seen that Ameghino refers this to the Miocene epoch: Lehmann-Nitsche speaks of it as Pliocene, Steinmann's opinion suggests a still later date, while Scott also declares that no greater age than that of the Pleistocene period can be assigned. The two specimens were obtained at very different

ALLUVIAL DEPOSITS AND CAVES

times, an interval of many years separating the dates of the respective discoveries. So far as is known, no mammalian or other animal remains have been yielded by the strata in question, so that the whole case in regard to evidence is one of the most unsatisfactory on record. Indeed the whole question of 'dating' the Argentine discoveries, whether absolutely or relatively, must be regarded as an unsolved problem.

Combe Capelle (*Dordogne*). The circumstances of this discovery were as follows. The skeleton lay in an extended position, and it had been placed in an excavation made for the purpose of interment. This excavation entered a stratum distinguished as Mousterian. But the interment is considered to be later, and of Aurignacian antiquity. Stone implements of Aurignacian type were disposed around the skeleton: in addition to these, a number of molluscan shells were arranged about the skull. This suggestion of ornament would of itself suggest the later period to which the skeleton is assigned. No remains of animals are mentioned in the accounts accessible to me.

Brüx (*Bohemia*). The Brüx skeleton was discovered in 1871. It lay some five feet beneath the surface in a deposit which seems to be an ancient one of fluviatile origin. The Biela river is not far from the spot. The bones were very fragmentary, and in particular the skull-cap has been reconstructed from no less than a dozen fragments. The limb bones were also fractured. Near the skeleton, some remains of an Ox were found on the same level. Two feet above the skeleton, a stone implement, seemingly a Neolithic axe, was brought to light.

The information is thus meagre in the extreme, and when the condition of the skull is taken into account, it is evident that the Brüx skeleton is not one upon which far-reaching arguments can be successfully based. The interest of the specimen depends above all upon the results of the careful analysis of its characters made by Professor Schwalbe[25] (1906).

Brünn (1871). This discovery was made at a depth of 4·5 metres in red löss. Close to the human bones lay the tusk and the shoulder-blade of a Mammoth. The same stratum subsequently yielded the skull of a young Rhinoceros (*R. tichorhinus*): some ribs of a Rhinoceros are scored or marked in a way suggestive of human activity: other ribs of the same kind were artificially perforated. More noteworthy, however, is a human figurine carved in ivory of a Mammoth tusk. Several hundreds of the shell of *Dentalium badense* lying close to the human remains were truncated in such a way as to suggest that they had once formed a necklace.

Galley Hill (*Kent*). The gravel-pit whence the skeleton was obtained invades the 'high-level terrace-gravel' of the Thames valley. Such is the opinion

III] ALLUVIAL DEPOSITS AND CAVES 83

of expert geologists (Hinton[26]). In the gravel-pit a section through ten feet of gravel is exposed above the chalk. The bones were eight feet from the top of the gravel. Palaeolithic implements of a primitive type have been obtained from the same deposit at Galley Hill. No precise designation seems to have been assigned to them. From the published figures, they seem to correspond to the earlier Acheulean or to the Chellean type. One in particular, resembles the implements found at Reculver, and I have recently seen similar specimens which had been obtained by dredging off the Kentish coast near Whitstable. Some of the Galley Hill implements are compared to the high plateau forms from Ightham. These must be of great antiquity. Professor Rutot in 1903 assigned the Galley Hill skeleton to a period by him named Mafflian. This diagnosis seems to have been based upon the characters of the implements. Recently however (1909) Professor Rutot has brought the skeleton down into the Strépyan epoch, which is much less ancient than that of Maffle.

The associated fauna comes now into consideration. From the Galley Hill gravel-pit no mammalian remains other than the human skeleton have been reported, but the fauna of the 'high-level terrace' has been ascertained by observations in the vicinity of Galley Hill as well as in other parts of the Thames basin. The mollusc *Cyrena fluminalis*, indicative of a sub-

tropical climate, has been found in these strata.
As regards the mammalian fauna, it is interesting
to compare the list given by Mr E. T. Newton in 1895,
with that published by Mr M. A. C. Hinton in 1910
on the basis of independent observations.

Mr Newton's list, 1895.

1. Elephas primigenius.
2. Hippopotamus.
3. Rhinoceros: species uncertain.
4. Bos.   ,,   ,,
5. Equus.   ,,   ,,
6. Cervus.   ,,   ,,
7. Felis leo.   ,,   ,,

Mr Hinton's list, 1910.

1. Elephas antiquus (a more primitive form than E. primigenius).
2. No Hippopotamus (this occurs later, in the Middle Terrace).
3. Rhinoceros megarhinus.
4. Bos : species uncertain.
5. Equus: species similar to the Pliocene E. stenonis.
6. Cervus: 3 species: one resembles the Fallow-deer (C. dama), a 'southern' form.
7. Felis leo.
8. Sus: species uncertain: bones of limbs shew primitive features.
9. Canis: species uncertain.
10. Delphinus: species uncertain.
11. Trogontherium: species differing from the Pliocene form.
12. Various smaller rodents, such as Voles.

No definitely 'Arctic' mammals are recorded: the general aspect of the above fauna shews a strong similarity to the Pliocene fauna, which appears to have persisted to this epoch without much alteration of the various types represented.

I	II	III
Classification by characters of human bones[1]	Example	Immediate surroundings
Division II Subdivision *B*	(1) Combe Capelle	Cave
,,	(2) Galley Hill	Alluvial drift of High Terrace[3]
,,	(3) Grimaldi (Mentone)	Cave
Subdivision *A*	(4) La Ferrassie	Cave
,,	(5) Pech de l'Aze	Cave
,,	(6) Le Moustier	Cave
,,	(7) La Chapelle	Cave
,,	(8) S. Brélade	Cave
,,	(9) Krapina	Cave (Rock-shelter)
,,	(10) Taubach	Alluvial Deposit[4]
Division I	(11) Mauer	Alluvial deposit
,,	(12) Trinil	Alluvial deposit

[1] South American remains and some others are omitted owing to insufficiency of
[2] Names of fossil varieties of Rhinoceros. These are very confused. The term and Cuvier. *R. merckii* represents *R. hemitœchus* of Falconer, which is the *R.*
[3] The formation of the High Terrace drift is earlier than the date of arrival into Britain from the south-east of Europe. But the Galley Hill man, if contemporary supposed by Klaatsch to be closely allied, and to have come into Europe through
[4] The upper strata at Taubach yielded Reindeer and Mammoth. Near Weimar,
[5] Typical Val d'Arno (Pliocene) form.

TABLE A

IV Circumstances and surroundings:	V	VI
Associated animals	Name of types of associated implements	
Reindeer	Aurignacian	Interment
Elephas antiquus Rhinoceros megarhinus[2] Trogontherium (Rodent) Mimomys (Rodent)	Acheulean to ?Strépyan	? No interment
Reindeer Hyaena spelaea Felis spelaea (Marmot in higher strata)	Mousterian ? also Aurignacian	Interment
Reindeer Bison priscus	Mousterian	Interment
Reindeer Bison priscus	Mousterian	(Head only found?)
Bos primigenius No reindeer	Mousterian	Interment
Reindeer (*scarce*) Bison priscus	Mousterian	Interment
Reindeer Bos ? sp. Rhinoceros tichorhinus	Mousterian	?
Rhinoceros merckii Cave Bear Bos primigenius Marmot (Arctomys)	Mousterian	
Elephas antiquus Rhinoceros merckii Felis leo No Hippopotamus	? Mousterian ? Upper Acheulean = Levallois ? Chellean	No interment
Elephas antiquus Rhinoceros etruscus[5] Ursus arvernensis No Hippopotamus	None found	No interment
Hippopotamus? Rhinoceros sivasondaicus Other Sivalik types	None found by Dubois	No interment

data relating to their surroundings.
R. *leptorhinus* should be avoided altogether. R. *megarhinus* represents the R. *leptorhinus* of Falconer
rhinus of Owen and Boyd Dawkins. R. *tichorhinus* is R. *antiquitatis* of Falconer and some German writers.
of the 'Siberian' invasion of Britain by certain Voles. Already in Pliocene times, some Voles had come
with the High Terrace drift, had arrived in Britain ages before the appearance of *Homo aurignacensis*
Central if not Northern Asia. The 'High Terrace' mammals have a 'Pliocene' facies.
Wüst says the stratigraphical positions of R. *merckii* and R. *antiquitatis* have been found inverted.

CHAPTER IV

ASSOCIATED ANIMALS AND IMPLEMENTS

THE most important of recent discoveries of the remains of early prehistoric man have now been considered. Not only the evidence of the actual remains, but also that furnished by their surroundings has been called upon. It is evident that the last decade has been remarkably productive of additions to the stock of information on these subjects.

In the next place, enquiry has to be made whether any relation exists between the two methods of grouping, viz. (1) that in which the characters of the skeletons are taken as the test, and (2) that dependent upon the nature of the surroundings. A first attempt to elucidate the matter can be made by means of a tabulated statement, such as that which follows.

In constructing this table, the various finds have been ordinated according to the degree of resemblance to modern Europeans presented by the respective skeletons. Thus Division II with Subdivision B heads the list. Then follows Subdivision A, and finally Division I will be found in the lowest place.

This order having been adopted, the remaining data were added in the sequence necessarily imposed upon them thereby.

(*a*) In an analysis of this table the several columns should be considered in order. Taking that headed 'Immediate surroundings,' it is evident that whereas most of the members of Division II were 'cave-men,' two exceptions occur. Of these, the Galley Hill skeleton is by far the most remarkable. The Taubach remains represent, it will be remembered, a form almost on the extreme confines of humanity. That it should resemble the members of Division I, themselves in a similar position, is not very remarkable. And indeed it is perhaps in accordance with expectation, that remains of the more remote and primitive examples should be discovered, so to speak, 'in the open.' All the more noteworthy therefore is the position of the Galley Hill man, whose place according to his surroundings is at the end of the list opposite to that assigned to him by his physical conformation.

(*b*) Passing to the 'Associated animals,' similar conclusions will be formed again. Thus in the first place, most of the 'cave-men' were accompanied by remains of the Reindeer. Le Moustier and Krapina are exceptions but provide Bison or Urus which are elsewhere associated with the Reindeer. Otherwise Galley Hill and Taubach again stand out

IV] ANIMALS AND IMPLEMENTS 87

as exceptions. Moreover they have again some features in common, just as has been noted in respect of their alluvial surroundings. For the Elephant (*E. antiquus*) is identical in both instances. But the Rhinoceros of the 'high level' terrace gravel is not the same as that found at Taubach, and though the succession is discussed later, it may be stated at once that the *Rhinoceros megarhinus* has been considered to stand in what may be termed a grand-parental relation to that of Taubach (*R. merckii*), the *Rhinoceros etruscus* of the Mauer Sands representing the intervening generation (Gaudry[27], 1888). For the various names, reference should be made to the list of synonyms appended to Table A. Should further evidence of the relative isolation of the Galley Hill skeleton be required, the gigantic beaver (Trogontherium) is there to provide it, since nowhere else in this list does this rodent appear. The paradoxical position of the Galley Hill skeleton having been indicated, it is convenient to deal with all the examples of skeletons from alluvial deposits taken as a single group, irrespective of their actual characters.

(i) The study of the animals found in the corresponding or identical *alluvial deposits*, leads to inferences which may be stated as follows. The Trinil (Java) fauna will not be included, since the Javanese and European animals are not directly comparable. If attention is confined to the remaining instances, viz. Galley Hill, Taubach and Mauer, agreement is shewn in respect of the presence of *Elephas antiquus*, and this is absent from all the cave-deposits considered here [*v. infra* (ii) p. 90]. A rhinoceros appears in all three localities, but is different in each. Finally, two (viz. Galley Hill and Mauer) of the three, provide at least one very remarkable mammalian form, viz. Trogontherium (*Mimomys cantianus* is equally suggestive) of the high-level gravels, and the *Ursus arvernensis* of the Mauer Sands.

The significance of these animals may be indicated more clearly by the following statement. If the history of *Elephas antiquus* be critically traced, this animal appears first in a somewhat hazy atmosphere, viz. that of the transition period between Pliocene and Pleistocene times. It is a more primitive form of elephant than the Mammoth. Indeed, Gaudry[27] (1888) placed it in a directly ancestral relation to the last-mentioned elephant. And though the two were contemporary for a space, yet *Elephas antiquus* was the first to disappear. Moreover this elephant has much more definite associations with the southern group of mammals than has the Mammoth. Its presence is therefore indicative of the considerable antiquity of the surrounding deposits, provided always that the latter be contemporaneous with it. With regard to the Rhinoceros, the species *R. megarhinus* and *R. etruscus* have been found in definitely Pliocene strata. The former (*R. megarhinus*) seems to have appeared earliest (at Montpellier), whereas the Etruscan form owes its name to the late Pliocene formations of the Val d'Arno, in which it was originally discovered. The third species (*R. merckii*) is somewhat later, but of similar age to *Elephas antiquus*, with which it constantly appears. It is remarkable that the *R. etruscus*, though not the earliest to appear, seems yet to have become extinct before the older *R. megarhinus*. The latter was contemporary with *R. merckii*, though it did not persist so long as that species. With regard to the three alluvial deposits, the Rhinoceros provides a means of distinction not indicated by the elephantine representative, and the presence of *R. etruscus* is a test for very ancient deposits. From what has been stated above, it follows that of the three localities the Mauer Sands have the more ancient facies, and it is significant that here also the human form proves to be furthest removed from modern men. But the other localities are not clearly differentiated, save that the Taubach strata are perhaps the more recent of the two.

Coming next to the 'peculiar' animals; the *Ursus arvernensis* of Mauer is almost as distinctively 'Pliocene' as its associate, *Rhinoceros etruscus*. The Taubach strata have yielded nothing comparable to these, nor to the Trogontherium (or Mimomys) of the high-level terrace gravel. These animals are also strongly suggestive of the Pliocene fauna.

To sum up, it will be found that the evidence of the Elephant is to the effect that these alluvial deposits are of early Pleistocene age. It leads to the expectation that the fauna in general will have a 'southern,' as contrasted with an 'arctic' aspect. From the study of the Rhinoceros it appears that the Mauer Sands are probably the most ancient in order of time, that the strata of Taubach are the latest of the three and that *Elephas antiquus* will occur there (as indeed it does).

The other animals mentioned clinch the evidence for the Pliocene resemblance, and (at latest) the early Pleistocene antiquity of the Mauer Sands and the high-level terrace gravels. Within the limits thus indicated, the deposit of Mauer is again shewn to be the oldest, followed by the terrace-gravels, while Taubach is the latest and youngest of the three. All the characteristic animals are now entirely extinct.

For the reasons stated above, the fossil Javanese mammals of Trinil have not been discussed. It will suffice to note that on the whole they indicate a still earlier period than those of the European deposits in question.

(ii) The animals associated with the *cave-men* now call for consideration. The great outstanding feature

IV] ANIMALS AND IMPLEMENTS 91

is the constancy with which the Reindeer is found. This leads to a presumption that the climate was at least temperate rather than 'southern.' Beyond this, it will be noted that in general the cave-fauna is more familiar in aspect, the Reindeer having survived up to the present day, though not in the same area. Again, save in one locality, not a single animal out of those discussed in connection with the alluvial deposits appears here. The exception is the Krapina rock-shelter. The surviving animal is *Rhinoceros merckii*, described above as one of the later arrivals in the epochs represented by the alluvial deposits. Krapina does not provide the Reindeer, and in this respect is contrasted again with the remaining localities. Yet the presence of the Marmot at Krapina may be nearly as significant as that of the Reindeer would be.

Another cave, viz. the Grotte des Enfants, may also need reconsideration. For instance, the *Rhinoceros merckii* was found in the deepest strata of this cave: but I do not consider that adequate evidence is given of its contemporaneity with the two human skeletons here considered. But the Reindeer is found in the same cave, as indicated in the table.

With the exception of Krapina therefore, the conditions are remarkably uniform. This conclusion is confirmed by the evidence from many caves not described in detail here because of the lack of human bones therein or the imperfection of such as were found. Such caves have yielded abundant evidence in regard to the 'associated fauna.' A few of the more important results of the investigation of the mammals may be given. Thus the distribution of the Reindeer is so constant that except in regard to its abundance or rarity when compared with the remains of the horse in the same cave, it is of little or no use as a discriminating agency. The Mammoth (*E. primigenius*) was contemporaneous with the Reindeer, but was plentiful while the Reindeer was still rare. A similar remark applies to the Hairy Rhinoceros (*R. tichorhinus*), and also to the Cave-Bear. The Cervidae (other than the Reindeer), the Equidae, the Suidae (Swine) and the smaller Rodentia (especially Voles) are under investigation, but the results are not applicable to the finer distinctions envisaged here.

To sum up the outcome of this criticism; it appears that of the cave-finds, Krapina stands out in contrast with the remainder, in the sense that its fauna is more ancient, and is indicative of a southern rather than a temperate environment. The latitude of Krapina has been invoked by way of explaining this difference, upon the supposition that the *Rhinoceros merckii* survived longer in the south. Yet Krapina does not differ in respect of latitude from the caves of Le Moustier and La Chapelle, while it is rather to the north of the Mentone

IV] ANIMALS AND IMPLEMENTS 93

caves. Lastly, some weight must be attached to the alleged discovery at Pont Newydd in Wales, of Mousterian implements with remains of *R. merckii*.

The fauna of the other caves suggests temperate, if not sub-arctic conditions of climate. In all cases, the cave-finds are assignable to a period later in time than that in which the fluviatile deposits (previously discussed) were formed. The cave-men thus come within the later subdivisions of the Pleistocene period.

(c) The fifth column of the table gives the types of stone implements found in association with the respective remains. As is well known, and as was stated in the introductory sentences of this book, stone artefacts constitute the second great class of evidence on the subject of human antiquity. As such they might appropriately have been accorded a separate chapter or even a volume. Here a brief sketch only of their significance in evidence will be attempted. The value of stone implements in deciding upon the age of deposits (whether in caves or elsewhere) depends upon the intimacy of the relation existing between various forms of implement and strata of different age. How close that intimacy really is, has been debated often and at great length. Opinions are still at variance in regard to details, but as to certain main points, no doubt remains. Yet the study is one in which even greater specialisation is needed than in respect of comparative osteology. The descriptions following these preliminary remarks are based upon as extensive an examination as possible, both of the literature, and of the materials.

To discuss the validity of the claims made in favour of or against the recognition of certain individual types will be impossible, save in the very briefest form. The better-known varieties have received names corresponding to the localities where they were first discovered, or where by reason of their abundance they led to the recognition of their special value as a means of classification. These designations will be employed without further definition or explanation, save in a few instances.

Commencing again with the fifth column of the table, the first point to notice is that no implements at all have been discovered in immediate association with the fossil remains at Mauer and Trinil (Java). Yet in the absence of evidence, it must not be concluded that the contemporary representatives of mankind were incapable of providing such testimony. Evidence will be adduced presently to shew the incorrectness of such a conclusion.

In the next place, the great majority of the cavemen are associated with implements of one and the same type, viz. the Mousterian, so called from the locality (Le Moustier) which has furnished so complete an example of ancient prehistoric man.

ANIMALS AND IMPLEMENTS

Lastly, the Galley Hill skeleton maintains the distinctive position assigned to it, for as in the previous columns, it disagrees also here with the majority of the examples ranged near it.

If enquiry be made as to the significance, *i.e.* the sequence in point of time and the general status of the various types of implements mentioned in the table, it will be found that all without exception are described as of Palaeolithic type. Indeed they furnish largely the justification for the application of that term (employed so often in Chapter II) to the various skeletons described there.

To these Palaeolithic implements, others of the Neolithic types succeeded in Europe. [It is necessary to insist upon this succession as European, since palaeoliths are still in use among savage tribes, such as the aboriginal (Bush) natives of South Africa.] Confining attention to palaeoliths and their varieties, the discovery of a form alleged to fill the gap separating the most ancient Neolithic from the least ancient Palaeolithic types may be mentioned. The implements were obtained from the cave known as Le Mas d'Azil in the south of France.

In Germany, the researches of Professor Schmidt[28] in the caverns of Württemburg have revealed a series of strata distinguished not only in position and sequence but also by the successive types of stone implements related to the several horizons. The sequence may be shewn most concisely if the deposits are compared in a tabular form as follows (Table I).

These caves give the information necessary for a correct appreciation of the position of all the cave-implements in Table A. Reverting to the latter, and having regard to the cave-men, both subdivisions of Division II (cf. Table A) appear, but no example or representative of the earliest form (designated by Division I). The fauna is entirely Pleistocene, if we except such a trifling claim to Pliocene antiquity as may be based upon the presence of *Rhinoceros merckii* at Krapina.

The results of this enquiry shew therefore that genuine Mousterian implements are of Pleistocene age, that they were fabricated by human beings of a comparatively low type, who lived in caves and were by occupation hunters of deer and other large ungulate animals. So much has long been known, but the extraordinary distinctness of the evidence of superposition shewn in Professor Schmidt's work at Sirgenstein, furnishes the final proof of results arrived at in earlier days by the slow comparison of several sites representing single epochs. That work also helps to re-establish the Aurignacian horizon and period as distinctive.

When attention is turned from the cave-finds to those in alluvial deposits, names more numerous but less familiar meet the view. As the animals have been shewn to differ, so the types of implements

TABLE I.

Levels	Type of Implement		Fauna
	Ofnet	Sirgenstein	
A. Most superficial	—	Bronze	
B. 1. Intermediate	Neolithic Azilian	—	
	Palaeolithic	—	
2. Deepest stratum at Ofnet	Magdalenian	Magdalenian	Myodes torquatus (the Banded Lemming)
3.	—	Solutréan	Fauna of a northern character throughout: with Reindeer, Mammoth, Rhinoceros tichorhinus and Horse
4.	—	Aurignacian	
5. Deepest stratum at Sirgenstein	—	Mousterian	Myodes obensis (a Siberian Lemming)

provide a marked contrast. Yet a transition is suggested by the claim made on behalf of Mousterian implements for the Taubach deposits, a claim which (it will be remembered) is absolutely rejected by some experts of high authority.

In pursuing the sequence of implements from the Mousterian back to still earlier types, cave-hunting will as a rule provide one step only, though this is of the greatest value. In a few caves, implements of the type made famous by discoveries in alluvial gravels at S. Acheul in France (and designated the Acheulean type) have been found in the deeper levels. Such a cave is that of La Ferrassie (cf. p. 74); another is that of La Chapelle, in which (it will be remembered) the Acheulean implements underlay the human interment. Kent's Hole in Devonshire is even more remarkable. For the lowest strata in this cavern yielded implements of the earliest Chellean form, though this important fact is not commonly recognised. Such caves are of the greatest interest, for they provide direct evidence of the succession of types, within certain limits. But the indefatigable labours of M. Commont[29] of Amiens have finally welded the two series, viz. the cave-implements and the river-drift implements, into continuity, by demonstrating in the alluvial deposits of the river Somme, a succession of types, from the Mousterian backwards to much more primitive forms. These newly-published results have been appropriately supplemented by

IV] ANIMALS AND IMPLEMENTS 99

discoveries in the alluvial strata of the Danube. Combining these results from the river deposits, and for the sake of comparison, adding those from the caves at Ofnet and Sirgenstein, a tabulated statement (Table II) has been drawn up.

The two examples of human skeletons from alluvial deposits given in Table A are thus assigned to epochs distinguished by forms of implement more primitive than those found usually in caves; and moreover the more primitive implements are actually shewn to occur in deeper (*i.e.* more ancient) horizons where superposition has been observed. The greater antiquity of the two river-drift men (as contrasted with the cave-men) has been indicated already by the associated animals, and this evidence is now confirmed by the characters of the implements.

It may be remarked again that the details of stratigraphical succession have but recently received complete demonstration, mainly through the researches of Messrs Commont, Obermaier[30], and Bayer[30]. The importance of such results is extraordinarily far-reaching, since a means is provided hereby of correlating archaeological with geological evidence to an extent previously unattained.

(*d*) It will be noted that this advance has taken little or no account of actual human remains. For in the nature of things, implements will be preserved in river deposits, where skeletons would quickly disintegrate and vanish.

TABLE II.

Type of Implement	A. Caves[1]			B. Alluvial deposits		
	Ofnet[2]	Sirgenst..n[2]	S. Acheul (Tellier)[3]	Willendorf (Austria)[4]	S. Acheul (Tellier, etc.)[3]
Neolithic 1.		Bronze	—	—	—
2.	Neolithic	—	—	—	—
Intermediate 3.	Azilian	—	—	—	—
Palaeolithic 4.	Magdalenian	Magdalenian	Magdalenian	—	—
5.	—	Solutréan	—	Solutréan	—
6.	—	Aurignacian	—	Aurignacian	—
7.	—	Mousterian	—	—	Mousterian
8.	—	—	—	—	Acheulean
9.	—	—	—	—	Chellean
10.	—	—	—	—	"Industrie grossière"

[1] For the occurrence of Achenlean and Chellean implements in caves, v. page 98.
[2] Schmidt, 1909.   [3] Commont, 1908.   [4] Obermaier and Bayer, 1909.

CH. IV] ANIMALS AND IMPLEMENTS 101

The next subject of enquiry is therefore that of the antiquity of Man as indicated by the occurrence of his artefacts.

The succession of Palaeolithic implements has just been given and discussed, as far back as the period marked by the Chellean implements of the lower river gravels (not necessarily the lower terrace) of S. Acheul. For up to this point the testimony of human remains can be called in evidence. And as regards the associated animals, the Chellean implements (Taubach) have been shewn to accompany a group of animals suggestive of the Pliocene fauna which they followed.

But implements of the type of Chelles have been found with a more definitely 'Pliocene' form of elephant than those already mentioned. At S. Prest and at Tilloux in France, Chellean implements are associated with *Elephas meridionalis*, a species destined to become extinct in very early Pleistocene times. Near the Jalón river in Aragon, similar implements accompany remains of an elephant described as a variety of *E. antiquus* distinctly approaching *E. meridionalis*.

In pursuing the evidence of human antiquity furnished by implements, a start may be made from the data corresponding to the Galley Hill skeleton in column 5 of Table A. Two divergent views are expressed here, since the alternatives "Acheulean" or "Strépyan" are offered in the table. In the former instance (Acheulean) a recent writer (Mr Hinton,

1910) insists on the Pliocene affinities of the high-level terrace mammals. But as a paradox, he states that the high-level terrace deposits provide implements of the Acheulean type, whereas the Chellean type would be expected, since on the Continent implements associated with a fauna of Pliocene aspect, are of Chellean type. To follow Mr Hinton in his able discussion of this paradox is tempting, but not permissible here; it must suffice to state that the difficulty is reduced if Professor Rutot's[31] view be accepted. For the Strépyan form of implement (which M. Rutot recognises in this horizon) is older than the others mentioned and resembles the Chellean type. To appreciate this, the sequence which Professor Rutot claims to have established is here appended.

A. *Pleistocene Period.*

(All Palaeolithic types except No. 1.)

1. Azilian
2. Magdalenian
3. Solutréan
4. Aurignacian
5. Mousterian

Types found in caves as well as in alluvial deposits.

6. Acheulean. Fauna of S.-E. Britain has a Pliocene aspect. High-level terrace of Thames valley (Hinton, 1910).
7. Chellean. Fauna of Continent has Pliocene affinities (Hinton, 1910).
8. Strépyan. Galley Hill Skeleton. High-level terrace, Thames basin (Rutot, 1911).
9. Mesvinian. Implements on surface of chalk-plateau, Ightham, Kent (Rutot, 1900).

ANIMALS AND IMPLEMENTS

10. Mafflian. Galley Hill skeleton (Rutot, 1903). Mauer jaw (Rutot, 1911).
11. Reutelian. High-level terrace of Thames basin, Rutot, 1900. The Reutelian implement is "eolithic," and is found unchanged in stages assigned to the Pliocene, Miocene and Oligocene periods (Rutot, 1911).

The duration of the Pleistocene period is estimated at about 139,000 years (Rutot, 1904).

B. *Pliocene Period.*

12. Kentian (Reutelian).

C. *Miocene Period.*

13. Cantalian (Reutelian).

D. *Oligocene Period.*

14. Fagnian (Reutelian).

E. *Eocene Period.*

15. [Eoliths of Duan and other French sites: not definitely recognised in 1911 by Rutot.]

Several results of vast importance would follow, should the tabulated suggestions be accepted unreservedly in their entirety.

An inference of immediate interest is to the effect that if Professor Rutot's view be adopted, the high-level terrace of the Thames valley is not contrasted so strongly with continental deposits containing the same mammals, as Mr Hinton suggests. For Professor Rutot's Strépyan period is earlier than the Chellean. It may be questioned whether Mr Hinton is right in assigning only Acheulean implements to the high-terrace gravels. Indeed Mr E. T. Newton (1895) expressly records the occurrence at Galley Hill, of implements more primitive than those of Acheulean form, and 'similar to those found by Mr B. Harrison on the high plateau near Ightham,'—*i.e.* the Mesvinian type of Professor Rutot. A final decision is perhaps unattainable at present. But on the whole, the balance of evidence seems to go against Mr Hinton; though *per contra* it will not escape notice that since 1903, Professor Rutot has 'reduced' the Galley Hill skeleton from the Mafflian to the Strépyan stage, and it is therefore possible that further reduction may follow.

Leaving these problems of the Galley Hill implements and the Strépyan period, the Mesvinian and Mafflian types are described by Professor Rutot as representatives of yet older and more primitive stages in the evolution of these objects. As remarked above (Chapter III), the Mauer jaw is referred by Professor Rutot to the Mafflian (implement) period of the early Pleistocene age, though the grounds for so definite a statement are uncertain.

More primitive, and less shapely therefore, than the Mafflian implements, are the forms designated 'Reutelian.' They are referred to the dawn of the Quaternary or Pleistocene period. But with these

IV] ANIMALS AND IMPLEMENTS 105

the initial stage of evolution seems to be reached. Such 'eoliths,' as they have been termed, are only to be distinguished by experts, and even these are by no means agreed in regarding them as products of human industry. If judgment on this vital point be suspended for the moment, it will be seen that Professor Rutot's scheme carries this evidence of human existence far back into the antiquity denoted by the lapse of the Pliocene and Miocene periods of geological chronology. But let it be remarked that when the names Kentian, Cantalian and Fagnian are employed, no claim is made or implied that three distinctive types of implement are distinguished, for in respect of form they are all Reutelian.

Herein the work of M. Commont must be contrasted with that of Professor Rutot. For the gist of M. Commont's researches lies in the demonstration of a succession of types from the more perfect to the less finished, arranged in correspondence with the superimposed strata of a single locality. A vertical succession of implements accompanies a similar sequence of strata.

Professor Rutot examines the Pliocene deposits in England, Miocene in France and Oligocene in Belgium, and finds the same Reutelian type in all. The names Kentian, Cantalian, and Fagnian should therefore be abandoned, for they are only synonyms for Pliocene-Reutelian, etc.

It is hard to gain an idea of the enormous duration of human existence thus suggested. But a diagram (Fig. 24) constructed by Professor Penck[32] is appended with a view to the graphic illustration of this subject. The years that have elapsed since the commencement of the Oligocene period must be numbered by millions. The human type would be shewn thus not merely to have survived the Hipparion, Mastodon and Deinotherium but to have witnessed their evolution and the parental forms whence they arose.

Such is the principal outcome of the opinions embodied in the tabulation of Professor Rutot. That observer is not isolated in his views, though doubtless their most energetic advocate at the present day. We must admire the industry which has conferred upon this subject the support of evidence neither scanty in amount, nor negligible in weight. But the court is still sitting, no final verdict being yet within sight.

While the so-called Eocene eoliths of Duan (Eure-et-Loire) fail to receive acceptance (Laville[33], 1906), even at Professor Rutot's hands (1911), it is otherwise with those ascribed to the Oligocene period. Mr Moir[34] of Ipswich has lately recognised prepalaeoliths beneath the Suffolk Crag (Newbourn) at Ipswich resting on the underlying London Clay.

Some objections to the recognition of the so-called 'eoliths' as artefacts may now be considered.

Glacial Epoch	Pliocene	Miocene	Oligocene
1. ⋁⋁			
2. Palaeolichs KENT	Aurillac		Boncelles
3. Equus.	Hipparion	Anchitherium	Palaeotherium
4. Elephas.	Mastodon	Deinotherium	

Fig. 24. Chart of the relative duration of Miocene, Pliocene and Pleistocene time: (From Penck.)

1. Line of oscillation of level of lowest snow-line. (Central Europe.)
2. Localities where 'eolithic implements' occur.
3. Names of representatives of ancestral forms of the modern Horse. The claim of Anchitherium to occupy the position it holds here, is strongly criticised by Depéret.
4. Names of representatives of ancestral forms of modern Elephants.

The chart is to be read from right to left. The gradual sinking of the snow-line is to be noticed, and the oscillations of the same line during the Glacial Period are also shewn (cf. Fig. 25).

(1) The case of the opponents rests mainly on a fourfold basis of argument. Thus the nature of the splintering or chipping is called in question. Some writers appeal to weathering, others to movements in the deposits ('earth-creep,' and 'foundering of drifts,' Warren[35] 1905, and Breuil, 1910), and others again to the concussions experienced by flints in a torrential rush of water. The last explanation is supported by observations on the forms of flints removed from certain rotary machines used in cement-factories (Boule[36], 1905).

(2) A second line of opposition impugns the association of the flints with the strata wherein they were found, or the geological age of those strata may be called in question as having been assigned to too early a period.

(3) Then (in the third place) comes the objection that the eoliths carry Man's existence too far back; having regard to the general development of the larger mammals, Pliocene Man might be accepted, but 'Oligocene' Man is considered incredible. Moreover the period of time which has elapsed since the Oligocene period must be of enormous length.

(4) In the last place will be mentioned criticism of the distribution of the eolithic type (Obermaier[37], 1908).

(1) Having regard to the first of these arguments, the balance of evidence appears so even and

IV] ANIMALS AND IMPLEMENTS

level that it is hardly possible to enter judgment on this alone. But experiments recently carried out by Mr Moir, and in Belgium by Munck and Ghilain (1907; cf. Grist[38], 1910) should do much to settle this point.

Moreover the 'wash-tub' observations in cement-factories (Boule, 1905) prove too much, for it is alleged that among the flint-refuse, fragments resembling Magdalenian or even Neolithic implements were found. Yet such forms are not recorded in association with the comparatively shapeless eoliths. Further experiments are desirable, but so far they support Professor Rutot and his school rather than their opponents.

(2) The position of the eoliths and the accuracy with which their immediate surroundings are determined may be impugned in some instances, but this does not apply to Mr Moir's finds at Ipswich, nor to the Pliocene eoliths found by Mr Grist[38] at Dewlish (1910).

(3) While the general evidence of palaeontology may be admitted as adverse to the existence of so highly-evolved a mammal as Man in the earlier Tertiary epochs, yet the objection is of the negative order and for this reason it must be discounted to some extent. If the lapse of time be objected to, Dr Sturge[39] (1909) is ready to adduce evidence of glacial action upon even Neolithic flints, and to propose a base-line for the commencement of the Neolithic phase no less than 300,000 years ago.

(4) The distribution of the implements finds a weak spot in the defences of the eolithic partisans. It is alleged that eoliths are almost always flints: and that they occur with and among other flints, and but rarely elsewhere. Palaeoliths (of flint) also occur among other flints, but they are not thus limited in their association. This distinction is admitted by some at least of the supporters of the 'artefact' nature of the eoliths, and the admission certainly weakens their case.

The question is thus far from the point of settlement, and it may well continue to induce research and discussion for years to come. That a final settlement for the very earliest stages is practically unattainable will be conceded, when the earliest conditions are recalled in imagination. For when a human being first employed stones as implements, natural forms with sharp points or edges would be probably selected. The first early attempts to improvise these or to restore a blunted point or edge would be so erratic as to be indistinguishable (in the result) from the effects of fortuitous collisions. While such considerations are legitimately applicable to human artefacts of Oligocene or Miocene antiquity, they might well appear to be less effective when directed to the Pleistocene representatives where signs of progress might be expected. Yet Professor Rutot (1911) does not distinguish even the Pleistocene

Reutelian from the Oligocene (eolithic) forms. If, on such evidence as this, early Pleistocene Man be recognised, Oligocene Man must needs be accepted likewise. Professor Rutot's mode of escape from this difficult position is interesting and instructive, if not convincing. It is effected by way of the assumption that in regard to his handiwork, Man (some say a toolmaking precursor of Man) was in a state of stagnation throughout the ages which witnessed the rise and fall of whole genera of other mammals. That this proposition is untrue, can never be demonstrated. On the other hand, the proposition may be true, and therefore the unprejudiced will maintain an open mind, pending the advent of more conclusive evidence than has been adduced hitherto.

CHAPTER V

HUMAN FOSSILS AND GEOLOGICAL CHRONOLOGY

In the preceding Chapter, the remains of Palaeolithic Man were studied in relation to the associated animals (especially mammals), and again (so far as possible) in connection with the accompanying implements. In the comparison of the different types of implement, evidence was adduced to shew that certain forms of these are distinctive of corresponding geological horizons. Of the three series, (1) human remains, (2) mammalian remains, (3) stone implements, the first two, (1) and (2), have been compared as well as (1) and (3). A comparison between (2) and (3) has now to be instituted. And this is of interest, for mammalian remains have been found in the presence of implements where no human bones could be discovered. Moreover the expectation is well founded, whereby the mammalian fauna will prove to supply information unobtainable from either human skeletons or implements by themselves. That information will bear upon the climatic conditions of the different phases which mark the geological history of Man. And in this way, a more perfect correlation

CH. V]   HUMAN FOSSILS   113

of the past history of Man with the later geological history of the earth may be fairly anticipated.

In Chapter IV, use was frequently made of the expression 'southern,' 'temperate' or 'sub-arctic,' in connection with the various groups of mammals mentioned in Table A. And while the geological period is limited, during which these investigations are profitably applicable, yet the matter is one of no small importance. For the very fact that the fauna can be described in one case as 'southern' in character, in another as 'temperate,' suggests some variation of climate. And the relation of the history of Man to the great variation of climate implied in the expression 'Glacial Period,' may be reasonably expected to receive some elucidation from this branch of study. It will be noticed that Man himself is at present comparatively independent of climate, and even in earlier times he was probably less affected than some other animals. But while the importance of these studies must be recognised, it is also very necessary to notice that as elsewhere so here the difficulties are great, and pitfalls numerous.

It is no part of the present work to attempt a history of the stages through which opinion passed in developing the conception embodied in the phrase 'Ice-Age.' Long before that idea had been formulated, the presence of animal remains both in cave and alluvial deposits was a matter of common knowledge. The late Professor Phillips is believed to have been the first to make definite use of the terms 'pre-glacial' and 'post-glacial' in reference to the later geological formations (1855). And to the pre-glacial era that geologist referred most of the ossiferous caves and fissures.

But in 1860, this, the accepted view, was overthrown by the late Dr Falconer[40] at least so far as the caves (with the exception of the Victoria Cave) then explored in Britain were concerned. In the same year, the post-glacial position and antiquity of various brick-earths and gravels of the Thames valley were considered to have been definitely established by the late Professor Prestwich. It is very important to note in this connection, that the palaeontological evidence of those brick-earths was nevertheless held to indicate pre-glacial antiquity and thus to contradict the evidence of stratigraphy. The method employed in the latter mode of enquiry consisted in ascertaining the relation of the boulder-clay to certain deposits distinguished by their fauna, the Mollusca being especially employed in the identifications. Boulder-clay seems, in this country, to have been taken as the premier indication of the glacial period; it was supposed to be a submarine deposit formed during a submergence of large parts of these islands in the course of that period. That the late Sir Charles Lyell dwelt upon the problems of the boulder-clay should also be

HUMAN FOSSILS

recalled, for he expressly recounts how constantly it proved a barrier marking the extreme limit to which the works of Man could be traced. Implements or even bones had been found in the drift and above the boulder-clay, but not below.

For a while no attempt seems to have been made to subdivide the boulder-clay or to question its exact identity over all the area occupied by it. Yet such a subdivision might have resulted in explaining the contradiction or paradox (curiously analogous to that propounded by Mr Hinton in 1910, cf. p. 102 supra) just mentioned as existing between the age to be assigned to the Thames river-drift upon (*a*) stratigraphical evidence ('post-glacial'), and (*b*) palaeontological evidence ('pre-glacial').

That there might be several deposits of the boulder-clay with intervening strata, does not appear to have been suggested. The Glacial period was long regarded as one and indivisible. By some able geologists that view is still held.

Yet even in those comparatively early days, some succession of glaciations was suspected. In 1845, Ramsay recognised three phases of ice-action in North Wales. In 1855, Morlot took in hand the work of charting the extent of several Swiss glaciations. At last the possibility of a subdivision of the boulder-clay was realised, and it was demonstrated by the researches of Sir A. Geikie[41] (1863). But such division of the boulder-clay leads directly to an inference of successive periods of deposition—and when the earlier opinion (whereby the boulder-clay was regarded as a submarine deposit) was partly abandoned in favour of its origin as a 'ground-moraine,' the plurality of glaciations was still more strongly supported. The work of Julien (Auvergne, 1869) and Professor James Geikie (1873) carries the story on to the year 1878 which is marked by a very memorable contribution from Professor Skertchley[42], by whom account was taken of the stratigraphical position of stone implements. The names of these pioneers (and that of Croll should be added to the list) may be fittingly recalled now that the names of later continental observers figure so largely. But the work of Professors Penck, Brückner, Boule and Obermaier, admirable as it is, may be regarded justly as an extension or amplification of pre-existing research.

A multiplicity of glaciations demonstrated whether by successive 'end-moraines,' or by a series of boulder-clays or 'tills,' implies intervening 'inter-glacial' epochs. To the earlier-recognised pre-glacial and post-glacial periods, one or more inter-glacial phases must therefore be added. Consequently the absence of evidence (indicative of Man's existence) from the boulder-clay need not exclude his presence in the inter-glacial deposits; and in fact the appearance of strongly-supported evidence that some implements of only Neolithic antiquity occur in inter-glacial surroundings, has been mentioned already (Chapter IV, Sturge, 1909). And thus, whether the series be one of grand oscillations constituting as many periods, or on the other hand a sequence of variations too slight to deserve distinctive terms, the fact of alternations prolonged over a considerable time seems to be established. Attempts to correlate various phases in the history of the animal and particularly of the human inhabitants of the affected area with these changes, still remained to be made.

Of such attempts, an early one, if not absolutely the earliest, stands to the credit of Dr Skertchley (1878). But in 1888 a much more definite advance was made by Professor Boule[43]. Still later came the suggestions of Professors Mortillet, Hoernes[44] (1903), Penck, Obermaier[45] (1909) and Tornqvist. And the employment of implements in evidence was found practicable by them. Ample compensation is thus provided for the lack of human bones, a deficiency almost as deplorable in 1911 as it was when Lyell called attention to it in 1863.

But the literature on this subject is so controversial and has attained such proportions, that the attempt to present current views will be limited to the discussion of the appended table (B). Here an endeavour has been made to submit the views expressed by the most competent observers of the day. The first point to which attention is directed consists in the manner in which the several glacial periods are distributed over the geological time-table. Boule claims one glaciation of Pliocene antiquity, followed by two Pleistocene glaciations. The remaining authors agree in ascribing all the glaciations to the Pleistocene period. Herein they follow the lead of Professor Penck, whose diagram of the oscillations in level of the snow-line in Central Europe is reproduced in Fig. 25. In the next place, the fact that Professor Penck's scheme was primarily intended to serve for the Swiss Alps must not be overlooked. That this system should leave traces everywhere else in Europe is not necessarily implied in accepting the scheme just mentioned.

In attempting to adjust the scale of glacial periods to that provided by the succession of implement-forms, it is suggested that a commencement should be made by considering the period designated Mousterian. If the position of the Mousterian period can be correlated with a definite subdivision of the Ice Age, then other periods will fall into line almost mechanically.

The first enquiry to make is that indicated in the introductory paragraphs of this Chapter, viz. what is the general nature of the fauna accompanying Mousterian implements? Investigation of the records shews that this is characteristically of a northern or a temperate, but not a southern type. For the

List of

Penck's scheme[1]	1908 Boule[2]	1908 Penck
Postglacial 4 = with Achen and other oscillations (Penck)	Magdalenian Solutréan[4]	Magdalenian
Glacial IV 2nd Pleistocene[2] Glaciation of Boule. "Würmian" of Penck	Mousterian	Solutréan[4]
Interglacial 3 = Riss-Würm interval (Penck)	Acheulean (Obermaier) Chellean	Mousterian (warm phase)
Glacial III 1st Pleistocene Glaciation of Boule. "Rissian" of Penck	Chellean	Mousterian (cold phase)
Interglacial 2 = Mindel-Riss interval (Penck)	?	Acheulean Chellean
Glacial II "Mindelian" of Penck	?	?
Interglacial 1 = Günz-Mindel interval (Penck)	?	?
Glacial I "Günzian" of Penck	?	?

[1] Penck postulates four glaciations, all "pleistocene."
[2] Boule recognises two pleistocene glaciations (seemingly Nos. III
[3] Skertchley's scheme is now ignored, if not abandoned, by the best is speculative.
[4] The differences between the rival schemes of Boule, Penck and its divisions are not indicated in this Table.

TABLE B.

types of associated implements.

1903	1905	1908	1878
Hoernes	Rutot	Sollas	Skertchley[3]
—	Neolithic period	?	Neolithic period
—	Lower Magdalenian Solutréan Aurignacian	?	Hessle Boulder-clay
Magdalenian	Mousterian Upper Acheulean	Acheulean	Palaeoliths of the "modern-valley" type. Valley-gravels of present Ouse, Cam, etc.
—	Lower Acheulean Chellean	[Chalky Boulder-clay of Hoxne]	Purple Boulder-clay
Solutréan	Strépyan Mesvinian Mafflean	?	Palaeoliths of "ancient-valley" type. ?Flood-gravels. Valleys do not correspond to modern rivers
—	—	?	Chalky Boulder-clay
Mousterian Chellean	—	?	Brandon beds with implements
—	—	?	Cromer Till. Later than Forest-Bed

IV of Penck), and one pliocene glaciation. The latter is not indicated in the Table.
horities. It has been introduced here on account of its historical interest only. Its correlation with the other schemes
rnes are best realised by comparing the position assigned to the Solutréan industry by each in turn. The löss and

Fig. 25. Chart of the oscillations of the snow-level in Central Europe during the Pleistocene period. (From Penck.)

In the uppermost space, *N* Neolithic Age. *Ma* Magdalenian. *Sol* Solutréan. *Günz, Mindel, Riss, Würm* denote the several glacial phases.

This chart is to be read from right to left; on the extreme right the snow-line is first shewn 300 m. above its present level. Then it falls to nearly 1200 m. below the present level, the fall corresponding to the Günzian glaciation. After this it nearly attains its former level, but does not quite reach the line marked + 300. This chart represents the part marked Glacial Epoch in Fig. 24, with which it should be compared.

combination commonly regarded as indicative of the southern type (viz. *Elephas antiquus, Rhinoceros merckii,* and *Hippopotamus major*) is very doubtfully demonstrable in this association, save in the very remarkable instance of the Grotte du Prince, Mentone, and Boule (1906) makes somewhat laboured efforts to explain this example, which is exceptional in his opinion. On the other hand, that combination does occur in well-recognised inter-glacial deposits, *e.g.* the Swiss Lignites of Dürnten, etc.

The Mousterian implements commonly accompany much more definitely northern animal forms, so that a glacial rather than an inter-glacial age is indicated. But there are four such glacial phases from which to choose in Professor Penck's scheme, and in Professor Boule's scheme there are two (for the 'Pliocene glaciation,' appearing in the latter, is hardly in question).

It will be seen (by reference to Table B) that Professor Boule assigns typical Mousterian implements to the most recent glacial period (Boule's No. III = Penck's No. IV = Würm), whereas Professor Penck places them in his penultimate grand period (Riss), carrying them down into the succeeding (Riss-Würmian) inter-glacial period.

Much diligence has been shewn in the various attempts to decide between these, the two great alternatives. (The view of Professor Hoernes, who assigns the Mousterian types to the first inter-glacial period of Penck, has received so little support as to render it negligible here.)

Upon an examination of the controversial literature, the award here given is in favour of Professor Boule's scheme. The following reasons for this decision deserve mention.

(1) Almost the only point of accord between the rival schools of thought, consists in the recognition by each side that the Magdalenian culture is post-glacial. But beyond this, the two factions seem to agree that the Mousterian culture is 'centred' on a glacial period but that it probably began somewhat earlier and lasted rather longer than that glacial period, whichever it might be.

(2) The Chellean implements, which precede those of Mousterian type, are commonly associated with a fauna of southern affinities. This denotes an inter-glacial period. Therefore an inter-glacial period is indicated as having preceded the Mousterian age. But after the Mousterian age, none of the subsequent types are associated with a 'southern fauna.'

Indications are thus given, to the following effect. The Mousterian position is such that a distinct inter-glacial period should precede it, and no such definite inter-glacial period should follow it. The last glacial period alone satisfies these requirements. The Mousterian position therefore coincides with the last great glaciation, whether we term this the fourth (with Professor Penck), or the third, with Professor Boule.

(3) The Mousterian industry characterises a Palaeolithic settlement at Wildkirchli in Switzerland: the position of this is indicated with great accuracy to be just within the zone limited by the moraine of the last great glacial period (Penck's No. IV or Würmian). The associated fauna is alleged to indicate that the age is not post-Würmian, as might be supposed. This station at Wildkirchli probably represents the very earliest Mousterian culture, and its history dates from the last phase of the preceding (*i.e.* the Riss-Würm) inter-glacial period. But it belongs to Penck's glaciation No. IV, not to No. III.

(4) Discoveries of implements of pre-Mousterian (Acheulean) form in the neighbourhood of the Château de Bohun (Ain, Rhone Basin, France, 1889), and Conliège (Jura, 1908), are accompanied by stratigraphical evidence whereby they are referred to an inter-glacial period later than the Riss glaciation (Penck's No. IV, Boule's No. III).

The remaining arguments are directed against the position assigned by Professor Penck to the Mousterian implements.

(5) Professor Penck admits that the epoch of the Mousterian type was glacial, and he recognises

that it was preceded by a definitely inter-glacial epoch, with a southern fauna. But by selecting his No. III as the glacial period in question he is led to postulate a subsequent but warmer inter-glacial subdivision of the Mousterian period. The difficulty is to find convincing evidence of this post-Mousterian inter-glacial period, and of the corresponding 'southern' fauna. Professor Penck believes that the 'southern' animals returned. Professor Boule can find no post-Mousterian evidence of such a fauna. The constituent forms became extinct or migrated southwards, never to return. If this contention be true, and there is much in its favour, Professor Boule's view must be adopted.

To shew how far-reaching some of the discussions are, attention may be directed to the fact that in this particular argument, much turns upon the nature of the implements found with the 'southern fauna' at Taubach (*v. ante* Chapters II and III). If the implements are of Mousterian type, they support Professor Penck's view, for the 'warm Mousterian' sought by him will thus be found: but if the type is Chellean, the arguments of Professor Boule are notably reinforced.

(6) The position assigned to one stage in the series of implements will affect all the rest. Professor Penck's view has been attacked with vigour and also

with great effect, on account of the position he allots to the type of Solutré. The consensus of opinion regarding the position of Solutré (*i.e.* its typical implements) is very extensive and quite definite. In effect, the type of Solutré is assigned to the newer (*jüngerer*) löss deposits. But these are also widely recognised as entirely post-glacial. Moreover in the last few years, the excavations in these particular löss-deposits in Lower Austria have not only confirmed that opinion, but have also revealed there the presence of Aurignacian implements, which closely follow those of Mousterian type.

Professor Penck's scheme seems therefore to carry the Solutréan implements too far back. The attempt to overcome this objection by attributing an earlier (? inter-glacial) age to the special variety of löss in question, has not been attended with conspicuous success.

Such are the main considerations upon which the decision has been taken in favour of Professor Boule's chronological scale. But when such an authority as Professor Sollas[46] (1908) is undecided, an amateur must not attempt to ignore the difficulties to be met. And while it is expedient to arrive at a final judgment, yet, in these controversies, the tendency is very marked to allow theory to run too far ahead of fact. Facts of the following kind are hard to reconcile with the schemes just described. (i) A

Mousterian type of implement is recorded by Commont from the later (younger) löss of the third terrace at S. Acheul. According to the theory, the type of Solutré, and not of Le Moustier, should have occurred. (ii) In this country at least, an admixture of 'northern' and 'southern' animals in a single deposit, has been demonstrated not infrequently, as in Italy also (Torre della Scalea, Cosenza). (iii) Professor Boyd Dawkins[47] (1910) insists upon the occurrence of Chellean, Acheulean, and Mousterian implements in one and the same British river deposit.

Consequently the distinction of a northern from a southern fauna may yet prove to be destitute of sound foundations. Many years ago, Saporta pointed out instances of regions with a sub-tropical climate actually adjacent to glacial areas. This subject has fortunately now the advantage of the attention and criticism provided by such talented observers as Mr Hinton, Professor Laville, and Professor Schmidt.

A trustworthy scheme of the relative chronology of culture (as denoted by the forms of implements), of mammalian variation and evolution (as shewn by the fauna), and of great climatic oscillations has not yet been obtained, but it has not been shewn to be unattainable. Meanwhile the schemes outlined in Table B mark a very great advance upon their predecessors.

It may be of interest to note that Professor Penck believes that the several periods varied both in duration and in intensity. Their relative proportions are shewn in Professor Penck's diagram (Fig. 25). The smaller oscillations, following the close of the last great glaciation (Würmian), should be noticed.

CHAPTER VI

HUMAN EVOLUTION IN THE LIGHT OF RECENT DISCOVERIES

IN this, the concluding Chapter, account is taken of the bearing of the foregoing discoveries and discussions, in relation with the light which they throw on the story of human development.

A. Up to a certain point, the evidence is strikingly favourable to the hypothesis of human evolution. By this is meant the gradual development of the modern type of skeleton found in association with a large and active brain, capable of manifesting its activity in a great variety of ways. Most of the oldest human skeletons just described, differ from this type. Although a difference cannot be demonstrated in respect of cranial capacity, yet those older skeletons are usually distinguished by the heavier jaw and by stout curved limb-bones of such length as to indicate an almost dwarf stature. Still these indications, even though marking a more primitive status, point undeniably to human beings. Passing beyond these, a few fragments remain to suggest a still earlier stage in evolution. And with these at least we find ourselves definitely on the neutral ground between the territories of man and ape, though even here on the human side of that zone.

In the same way, and again up to a certain point, the characters of human implements confirm the inferences drawn from the skeleton. For the older implements are re-gressively more and more crude, and an increasing amount of skill is needed to distinguish artefact from natural object.

Again, the associated animals seem to become less familiar, and the percentage of extinct species increases the further we peer into the stages of the past.

One of the most remarkable researches ever published upon these subjects is due to a group of scientists associated with Professor Berry of Melbourne University. In this place, only the most important of their memoirs (1910) can be called in evidence. In those particular publications, the initial objective was an attempt to measure the degree of resemblance between different types of skull. That endeavour may be roughly illustrated by reference to Fig. 26, in which tracings of various skull-outlines are adjusted to a conventional base-line. Should a vertical line be drawn from the mid-point of the base-line so as to cut the several contours, the vertical distances between the successive curves could be measured. The distance separating Pithecanthropus (*P.E.* of

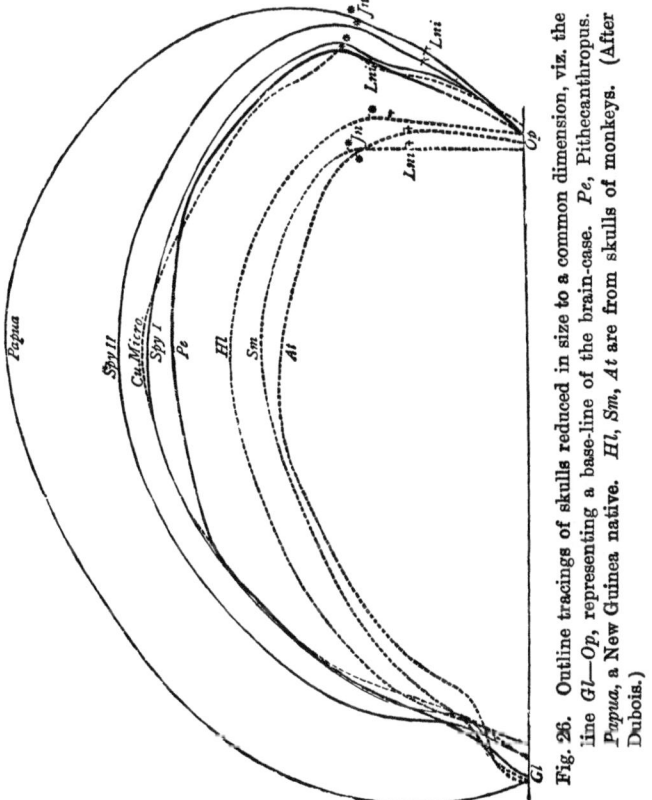

Fig. 26. Outline tracings of skulls reduced in size to a common dimension, viz. the line $Gl-Op$, representing a base-line of the brain-case. Pe, Pithecanthropus. $Papua$, a New Guinea native. Hl, Sm, At are from skulls of monkeys. (After Dubois.)

the figure) from that of the corresponding curve for
the Spy skull No. 1 (Spy 1 of the figure) is clearly less
than the distance between the curves for the second
Spy skull (Spy 2) and the Papuan native.

But Mr Cross used a much more delicate method,
and arrived at results embodied in the figure (27)
reproduced from his memoir. A most graphic
demonstration of those results is provided in this
chart. Yet it must be added, that the Galley Hill
skull, although shewn in an intermediate position,
should almost certainly be nearer the upper limit.
This criticism is based upon the conviction that many
of the measurements upon which the results are
dependent, assign to the Galley Hill skull a lowlier
status than it originally possessed before it became
distorted (posthumously). Again the Pithecanthropus
is apparently nearer to the Anthropoid Apes than to
Mankind of to-day. Let it be noticed however that
this is not necessarily in contradiction with the opinion
expressed above (p. 128 line 2). For Mr Cross' diagram
is based upon cranial measurements, whereas the
characters of the thigh-bone of Pithecanthropus tend
to raise it in the general scale of appreciation. On
the whole then, the evolutionary hypothesis seems to
receive support from three independent sources of
evidence.

B. But if in one of the very earliest of those
stages, a human form is discovered wherein the

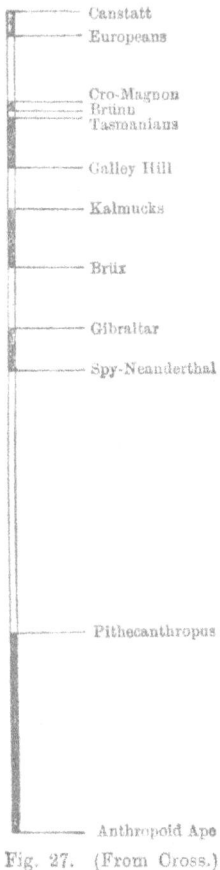

Fig. 27. (From Cross.)

characters of the modern higher type are almost if not completely realised, the story of evolution thus set forth receives a tremendous blow. Such has been the effect of the discovery of the Galley Hill skeleton. Time after time its position has been called 'abnormal' or 'isolated,' because it provides so many contrasts with the skeletons found in deposits regarded perhaps as leading towards but admittedly more recent than the Galley Hill gravel. And the juncture is long past at which its exact relation to that gravel could be so demonstrated as to satisfy the demands raised in a connection so vital to an important theory.

Some authors of great experience have refused to recognise in evidence any claim made on behalf of the Galley Hill skeleton. Yet it is at least pardonable to consider some of the aspects of the situation created by its acceptance.

(i) For instance, the argument is reasonable, which urges that if men of the Galley Hill type preceded in point of time the men of the lower Neanderthal type, the ancestry of the former (Galley Hill) must be sought at a far earlier period than that represented by the Galley Hill gravels. As to this, it may be noted that the extension of the 'human period,' suggested by eoliths for which Pliocene, Miocene, and even Oligocene antiquity is claimed, will provide more than this argument demands. The suggestion that a flint-chipping precursor of Man existed in Miocene time was made as long ago as 1878 by Gaudry[48].

(ii) But if this be so, the significance of the Neanderthal type of skeleton is profoundly altered. It is no longer possible to claim only an 'ancestral' position for that type in its relation to modern men. It may be regarded as a degenerate form. Should it be regarded as such, a probability exists that it ultimately became extinct, so that we should not expect to identify its descendants through many succeeding stages. That it did become extinct is a view to which the present writer inclines. Attempts have been made to associate with it the aborigines of Australia. But an examination of the evidence will lead (it is believed) to the inference that the appeal to the characters of those aborigines is of an illustrative nature only. Difficulties of a similar kind prevent its recognition either in the Eskimo, or in certain European types, although advocates of such claims are neither absent nor obscure.

Again, it is well to enquire whether any other evidence of degeneration exists in association with the men of the Neanderthal type. The only other possible source is that provided by the implements. This is dangerous ground, but the opinion must be expressed that there is some reason to believe that Mousterian implements (which rather than any other mark the presence of the Neanderthal type of skeleton) do present forms breaking the sequence of implement-evolution. One has but to examine the material, to become impressed with the inferiority of workmanship displayed in some Mousterian implements to that of the earlier Acheulean types. In any case, a line of evidence is indicated here, which is not to be overlooked in such discussions.

(iii) The Galley Hill skeleton has been described as comparatively isolated. Yet if it be accepted as a genuine representative of Man in the age of the gravel-deposits of the high-level terrace, it helps towards an understanding of the characters of some other examples. Thus a number of specimens (rejected by many authors as lacking adequate evidence of such vast antiquity as is here postulated) appear now, in this new light, as so many sign-posts pointing to a greater antiquity of that higher type of human skeleton than is usually recognised. Above all (to mention but a few examples), the cranium of Engis, with those from S. Acheul (discovered in 1861 by Mr H. Duckworth), and Tilbury, the fragment of a human skull from gravel at Bury St Edmunds, and a skeleton discovered near Ipswich beneath the boulder-clay in October 1911, seem to find their claims enhanced by the admission of those proffered on behalf of the Galley Hill specimen. And since Huxley wrote his memoir on the skulls from Engis and the Neanderthal, the significance of the former (Engis), fortified by the characters of the Galley Hill skeleton, has been greatly increased. Consequently it is not surprising to find confident appeals to the characters of a Galley Hill Race or Stock, near associates being the specimens mentioned in a preceding chapter as Brünn (1891) and the Aurignac man next to be considered. The relations of these to the well-known Cro-Magnon type will be mentioned in the next paragraph.

C. The appearance of the higher type of humanity in the period next following the Mousterian, viz. that distinguished by the Aurignacian type of implement, has now to be discussed. As already remarked, the man of Aurignac, as compared with him of the Neanderthal, has less protruding jaws, the lower jaw in particular being provided with the rudiment of a chin, while the limb bones are slender and altogether of the modern type. Upon such contrasts a remarkable theory has been based by Professor Klaatsch[49]. He made a comparison between the anthropoid apes on the one hand, and the two human types on the other (Fig. 28). As a result, he pointed out that the Orang-utan differs from the Gorilla much as the Aurignac does from the Neanderthal man. Assuming this statement to be correct, a hypothesis is elaborated to the effect that

136 PREHISTORIC MAN [CH.

two lines of human descent are here in evidence. Of these one includes an ancestor common to the Orang-utan (an Asiatic anthropoid ape) and the Aurignac man; the other is supposed to contain an

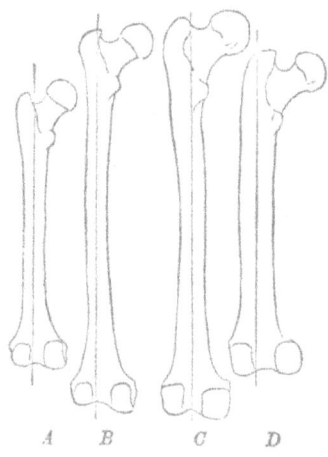

Fig. 28. Various thigh-bones arranged to shew the alleged similarity between *A* Orang utan and *B* Aurignac man, as also between *N* Neanderthal and *D* Gorilla. *A* and *B*, while resembling each other, are to be contrasted with *C* and *D*. They are referred to as the A/O and N/G groups. (From Klaatsch.)

ancestor common to the Gorilla (of African habitat), and the Neanderthal man.

The further development of the story includes the following propositions. The more primitive and

Gorilla-like Neanderthal type is introduced into Europe as an invader from Africa. Then (at a subsequent epoch probably) an Asiatic invasion followed. The new-comers owning descent from an Orang-utan-like forerunner are represented by the Aurignac skeleton and its congeners. In various respects they represented a higher type not only in conformation but in other directions. Having mingled with the Neanderthal tribes, whether by way of conquest or pacific penetration, a hybrid type resulted. Such was the origin of the Cro-magnon race.

The hypothesis has been severely handled, by none more trenchantly than by Professor Keith[50]. A notable weakness is exposed in the attribution to the ancestors of the Orang-utan so close an association to any human ancestral forms, as Professor Klaatsch demands. To those familiar with the general anatomy of the Orang-utan (*i.e.* the anatomy of parts other than the skeleton) the difficulties are very apparent.

Another effect of the hypothesis is that the so-called Neanderthaloid resemblances of the aborigines of Australia are very largely if not entirely subverted. This would not matter so much, but for the very decided stress laid by Professor Klaatsch upon the significance of those resemblances (cf. Klaatsch, 1909, p. 579, 'Die Neanderthalrasse besitzt zahlreiche australoide Anklänge'). Again in earlier days, Professor

Klaatsch supported a view whereby the Australian continent was claimed as the scene of initial stages in Man's evolution. Finally, up to the year 1908, Professor Klaatsch was amongst the foremost of those who demand absolute exclusion of the Orangutan and the Gorilla from any participation in the scheme of human ancestry.

Having regard to such facts and to such oscillations of opinion, it is not surprising that this recent attempt to demonstrate a 'diphyletic' or 'polyphyletic' mode of human descent should fail to convince most of those competent to pronounce upon its merits.

Yet with all its defects, this attempt must not be ignored. Crude as the present demonstration may be, the possibility of its survival in a modified form should be taken into account. These reflections (but not necessarily the theory) may be supported in various ways. By a curious coincidence, Professor Keith, in rebutting the whole hypothesis, makes a statement not irrelevant in this connexion. For he opines that 'the characters which separate these two types of men (viz. the Aurignac and Neanderthal types) are exactly of the same character and of the same degree as separate a blood-horse from a shire-stallion.' Now some zoologists have paid special attention to such differences, when engaged in attempts to elucidate the ancestry of the modern types of horse. As a result of their studies, Professors Cossar Ewart and Osborn (and Professor Ridgeway's name should be added to theirs) agree that proofs have been obtained of the 'multiple nature of horse evolution' (Osborn). If we pass to other but allied animals, we may notice that coarser and finer types of Hipparion (*H. crassum* and *H. gracile*) have been contrasted with each other. A step further brings us to the Peat-hog problem (*Torf-Schwein Frage* of German writers), and in the discussion of this the more leggy types of swine are contrasted with the more stocky forms. Owen (in 1846) relied on similar points for distinguishing the extinct species of Bovidae (Oxen) from one another. The contrast may be extended even to the Proboscidea, for Dr Leith Adams believed that the surest test of the limb bones of *E. antiquus* was their stoutness in comparison with those of *E. primigenius*. This is the very character relied upon by Professor Klaatsch in contrasting the corresponding parts of the human and ape skeletons concerned. But such analogies must not be pressed too far. They have been adduced only with a view to justifying the contention that the diphyletic scheme of Professor Klaatsch may yet be modified to such an extent as to receive support denied to it in its present form.

D. In commenting upon the hypothesis expounded by Professor Klaatsch, mention was made of its bearing upon the status of the Cro-Magnon race. This is but part of a wide subject, viz. the attempt to trace in descent certain modern European types. It is necessary to mention the elaborate series of memoirs now proceeding from the pen of Dr Schliz[51], who postulates four stocks at least as the parent forms of the mass of European populations of to-day. Of these four, the Neanderthal type is regarded as the most ancient. But it is not believed to have been extirpated. On the contrary its impress in modern Europe is still recognisable, veiled though it may be in combination with any of the remaining three. The latter are designated the Cro-Magnon, Engis, and Truchère-Grenelle types, the last-mentioned being broad-headed as contrasted with all the rest. Of Professor Schliz' work it is hard to express a final opinion, save that while its comprehensive scope (without excessive regard to craniometry as such) is a feature of great value, yet it appears to lack the force of criticism based upon extensive anatomical, *i.e.* osteological study.

E. The remarkable change in Professor Klaatsch's views on the part played by the anthropoid apes in human ancestral history has been already mentioned. In earlier days the Simiidae were literally set aside by Professor Klaatsch. But although the anthropoid monkeys have gained an adherent, they still find their claim to distinction most energetically combated by Professor Giuffrida-Ruggeri[52]. The latter declares that though he now (1911) repeats his views, it is but a repetition of such as he, following De Quatrefages, has long maintained. In this matter also, the last word will not be said for some time to come.

F. The significance of the peculiar characters of massiveness and cranial flattening as presented by the Neanderthal type of skeleton continues to stimulate research. In addition to the scattered remarks already made on these subjects, two recently-published views demand special notice.

(i) Professor Keith has (1911) been much impressed with the exuberance of bone-formation, and the parts it affects in the disease known as Acromegaly. The disease seems dependent upon an excessive activity of processes regulated by a glandular body in the floor of the brain-case (the pituitary gland). The suggestion is now advanced that a comparatively slight increase in activity might result in the production of such 'Neanderthaloid' characters as massive brow-ridges and limb bones. (Of existing races, some of the aborigines of Australia would appear to exemplify this process, but to a lesser degree than the extinct type, since the aboriginal limb bones are exempt.) Professor Keith adopts the view that the Neanderthal type is ancestral to the modern types. And his argument seems to run further to the following effect: that the evolution of the modern from the

Neanderthal type of man was consequent on a change in the activity of the pituitary gland.

It is quite possible that the agency to be considered in the next paragraph, viz. climatic environment, may play a part in influencing pituitary and other secretions. But heavy-browed skulls (and heavy brows are distinctive tests of the glandular activity under discussion) are not confined to particular latitudes, so that there are preliminary difficulties to be overcome in the further investigation of this point. It is possible that the glandular activity occasionally assumed pathological intensity even in prehistoric times. Thus a human skull with Leontiasis ossea was discovered near Rheims at a depth of fifteen feet below the level of the surrounding surface.

(ii) Dr Sera[58] (1910) has been led to pay particular attention to the remarkably flattened cranial vaulting so often mentioned in the preceding paragraphs. As a rule, this flattening has been regarded as representative of a stage in the evolution of a highly-developed type of human skull from a more lowly, in fact a more simian one. This conclusion is challenged by Dr Sera. The position adopted is that a flattened skull need not in every case owe its presence to such a condition as an early stage in evolution assigns to it. Environment, for which we may here read climatic conditions, is a possible and alternative influence.

If sufficient evidence can be adduced to shew that the flattened cranial arc in the Neanderthal skull does actually owe its origin to physiological factors through which environment acts, the status of that type of skull in the evolutionary sequence will be materially affected. A successful issue of the investigation will necessitate a thorough revision of all the results of Professor Schwalbe's work[54], which established the Neanderthal type as a distinct species (*Homo primigenius*) followed closely and not preceded by a type represented by the Gibraltar skull. Dr Sera commenced with a very minute examination of the Gibraltar (Forbes Quarry) skull. In particular, the characters of the face and the basal parts of the cranium were subjected to numerous and well-considered tests. As a first result of the comparison of the parts common to both crania, Dr Sera believes that he is in a position to draw correct inferences for the Neanderthal skull-cap in regard to portions absent from it but present in the Forbes Quarry skull.

But in the second place, Dr Sera concludes that the characters in question reveal the fact that of the two, the Gibraltar skull is quite distinctly the lowlier form. And the very important opinion is expressed that the Gibraltar skull offers the real characters of a human being caught as it were in a lowly stage of evolution beyond which the Neanderthal skull together with all others of its class have already passed. The final extension of these arguments is also of remarkable import. The Gibraltar skull is flattened owing to its low place in evolution. But as regards the flatness of the brain-case (called the platycephalic character) of the Neanderthal calvaria and its congeners (as contrasted with the Gibraltar specimen), Dr Sera suggests dependence upon the particular environment created by glacial conditions. The effect is almost pathological, at least the boundary-line between such physiological flattening and that due to pathological processes is hard to draw. Upon this account therefore, Dr Sera's researches have been considered here in close association with the doctrines of Professor Keith.

Dr Sera supports his argument by an appeal to existing conditions: he claims demonstration of the association (regarded by him as one of cause and effect) between arctic latitudes or climate on the one hand, and the flattening of the cranial vault on the other. Passing lightly over the Eskimo, although they stand in glaring contradiction to his view, he instances above all the Ostiak tribe of hyperborean Asia. The platycephalic character has a geographical distribution. Thus the skull is well arched in Northern Australia, but towards the south, in South Australia and Tasmania, the aboriginal skull is much less arched. It is thus shewn to become more distinctly platycephalic towards the antarctic regions, or at least in the regions of the Australian Continent considered by Professor Penck to have been glaciated. So too among the Bush natives of South Africa as contrasted with less southern types.

The demonstration of a latitudinal distribution in the New World is complicated by the presence of the great Cordillera of the Rocky Mountains and Andes. Great altitudes are held by Dr Sera to possess close analogy with arctic or antarctic latitudes. Therefore the presence of flat heads (artificial deformation being excluded) in equatorial Venezuela is not surprising.

It is felt that the foregoing statement, though made with every endeavour to secure accuracy, gives but an imperfect idea of the extent of Dr Sera's work. Yet in this place, nothing beyond the briefest summary is permissible. By way of criticism, it cannot be too strongly urged that the Eskimo provide a head-form exactly the converse of that postulated by Dr Sera as the outcome of 'glacial conditions.' Not that Dr Sera ignores this difficulty, but he brushes it aside with treatment which is inadequate. Moreover, the presence of the Aurignac man with a comparatively well-arched skull, following him of the Mousterian period, is also a difficulty. For the climate did not become suddenly cold at the end of the Mousterian period, and so far as evidence of arctic human surroundings goes, the fauna did not become less arctic in the Aurignac phase.

Conclusion.

In section A of this chapter, an outline was given of the mode in which the evolution of the human form appears to be traceable backwards through the Neanderthal type to still earlier stages in which the human characters are so elementary as to be recognisable only with difficulty.

Then (B) the considerations militating against unquestioning acquiescence in that view were grouped in sequence, commencing with the difficulties introduced by the acceptance (in all its significance) of the Galley Hill skeleton. From an entirely different point of view (C), it was shewn that many difficulties may be solved by the recognition of more than one primordial stock of human ancestors. Lastly (F) came the modifications of theory necessitated by appeals to the powerful influence of physiological factors, acting in some cases quite obscurely, in others having relation to climate and food.

The impossibility of summing up in favour of one comprehensive scheme will be acknowledged. More research is needed; the flatness of a cranial arc is but one of many characters awaiting research. At the present time a commencement is being made with regard to the shape and proportions of the cavity bounded by the skull. From such characters we may aspire to learn something of the brain which was once active within those walls. Yet to-day the researches of Professors Keith and Anthony provide little more than the outlines of a sketch to which the necessary details can only be added after protracted investigation.

It is tempting to look back to the time of the publication of Sir Charles Lyell's 'Antiquity of Man.' There we may find the author's vindication of his claims (made fifty years ago) for the greater antiquity of man. In comparison with that antiquity, Lyell believed the historical period 'would appear quite insignificant in duration.' As to the course of human evolution, it was possible even at that early date to quote Huxley's opinion 'that the primordial stock whence man has proceeded need no longer be sought ...in the newer tertiaries, but that they may be looked for in an epoch more distant from the age of the Elephas primigenius than that is from us.'

The human fossils at the disposal of those authors included the Neanderthal, the Engis, and the Denise bones. With the Neanderthal specimen we have (as already seen) to associate now a continually increasing number of examples. And (to mention the most recent discovery only) the Ipswich skeleton (p. 151) provides in its early surroundings a problem as hard to solve as those of the Engis skull and the 'fossil man of Denise.' But we have far more valuable evidence than Lyell and Huxley possessed, since the incomparable remains from Mauer and Trinil provide an interest as superior on the anatomical side as that claimed in Archaeology by the Sub-crag implements.

Turning once more to the subject of human remains, the evolution of educated opinion and the oscillations of the latter deserve a word of notice. For instance, in 1863, the Engis skull received its full and due share of attention. Then in a period marked by the discoveries at Spy and Trinil, the claims of the Engis fossil fell somewhat into abeyance. To-day we see them again and even more in evidence. So it has been with regard to details. At one period, the amount of brain contained within the skull of the Neanderthal man was underestimated. Then that opinion was exchanged for wonder at the disproportionately large amount of space provided for the brain in the man of La Chapelle. The tableau is changed again, and we think less of the Neanderthal type and of its lowly position (in evolutionary history). Our thoughts are turned to a much more extended period to be allotted to the evolution of the higher types. Adaptations to climatic influences, the possibilities of degeneracy, of varying degrees of physiological activity, of successful (though at first aberrant) mutations all demand attention in the present state of knowledge.

If progress since the foundations were laid by the giant workers of half a century ago appears slow and the advance negligible, let the extension of our recognition of such influences and possibilities be taken into account. The extraordinarily fruitful results of excavations during the last ten years may challenge comparison with those of any other period of similar duration.

APPENDIX

The forecast, made when the manuscript of the first impression of this little book was completed, and in reference to the rapid accumulation of evidence, has been justified.

While it would be impossible to provide a review of all the additional literature of the last few months, it is thought reasonable to append notes on two subjects mentioned previously only in the preface.

(A) A short account of the 'La Quina' skeleton has now appeared (in 'L'Anthropologie,' 1911, No. 6, p. 730).

The skull is of the form described so often above, as distinctive of the Neanderthaloid type, but the brow-ridges seem even more massive than in the other examples of that race. The cranial sutures are unclosed, so that the individual is shewn to be of mature age, or at any rate, not senile. The teeth are, however, much worn down. Nearly all the teeth have been preserved in situ, and they present certain features which have been observed in the teeth found in Jersey (S. Brélade's Cave).

The skeleton lay in a horizontal position, but no evidence of an interment has been adduced. The bones were less than a metre below the present surface, and in a fine mud-like deposit, apparently ancient, and of a river-bed type. Implements were also found, and are referred unhesitatingly to the same horizon as the bones. The Mousterian period is thus indicated, but no absolutely distinctive implements were found. The general stratigraphical conditions are considered to assign the deposit to the base of what is termed the 'inferior Mousterian' level.

APPENDIX 151

(B) The 'sub-boulder-clay' skeleton, discovered near Ipswich in 1911, was in an extraordinarily contracted attitude. Many parts are absent or imperfect, owing to the solvent action of the surroundings, but what remains is sufficient to reveal several features of importance (cf. fig. 29).

Save in one respect, the skeleton is not essentially different from those of the existing representatives of humanity. The exception is provided by the shin-bone. That of the right limb has been preserved, and it presents an anomaly unique in degree, if not in kind, viz.: the substitution of a rounded for a sharp or keel-like edge to the front of the bone. It can hardly be other than an individual peculiarity, though the Spy tibia (No. 1) suggests (by its sectional contour) the same conformation.

So far as the skeleton is concerned, even having regard to the anomaly just mentioned, there is no good reason for assigning the Ipswich specimen to a separate racial type.

Its interest depends largely upon the circumstances of its surroundings. It was placed beneath about four feet of 'boulder-clay,' embedded partly in this and, to a much smaller extent, in the underlying middle-glacial sand which the bones just entered.

There is some evidence that the surface on which the bones lay was at one time exposed as an old 'land-surface.' A thin band of carbonised vegetable matter (not far beneath the bones) contains the remains of land plants. On this surface the individual whose remains have been preserved is supposed to have met with his end, and to have been overwhelmed in a sand drift. The latter it must be supposed was then removed, to be replaced by the boulder-clay.

Several alternatives to this rather problematical interpretation could be suggested. The most obvious of these is that we have to deal here with a neolithic interment, in a grave of which the floor just reached the middle-glacial sand of the locality. If we enquire what assumptions are requisite for the adoption of this particular alternative, we shall find, I think, that they are not very different in degree from those which are entailed by the supposition that the skeleton is really that of 'sub-boulder-clay' man.

The contracted attitude of the skeleton, and our familiarity of this as a feature of neolithic interments, taken together with the fact that the skeleton does not differ essentially from such as occur in interments of that antiquity, are points in favour of the neolithic age of the specimen. On the other hand, Mr Moir would urge that man certainly existed in an age previous to the deposition of the boulder-clay; that the implements discovered in that stratum support this claim; that the recent discovery of the bones of a mammoth on the same horizon (though not in the immediate vicinity) provides further support; that the state of mineralisation of the bones was the same in both cases, and that it is at least significant that they should be found on strata shewn (by other evidence) to have once formed a 'land-surface.'

On the whole then, the view adopted here is, that the onus of proof rests at present rather with those who, rejecting these claims to the greater antiquity of this skeleton, assign it to a far later date than that to which even the overlying Boulder-clay is referred. And, so far as the literature is at present available, the rejection does not seem to have been achieved with a convincing amount of certainty.

It is to be remarked, finally, that this discovery is entirely distinct from those made previously by Mr Moir in the deposits beneath the Red Crag of Suffolk, with which his name has become associated.

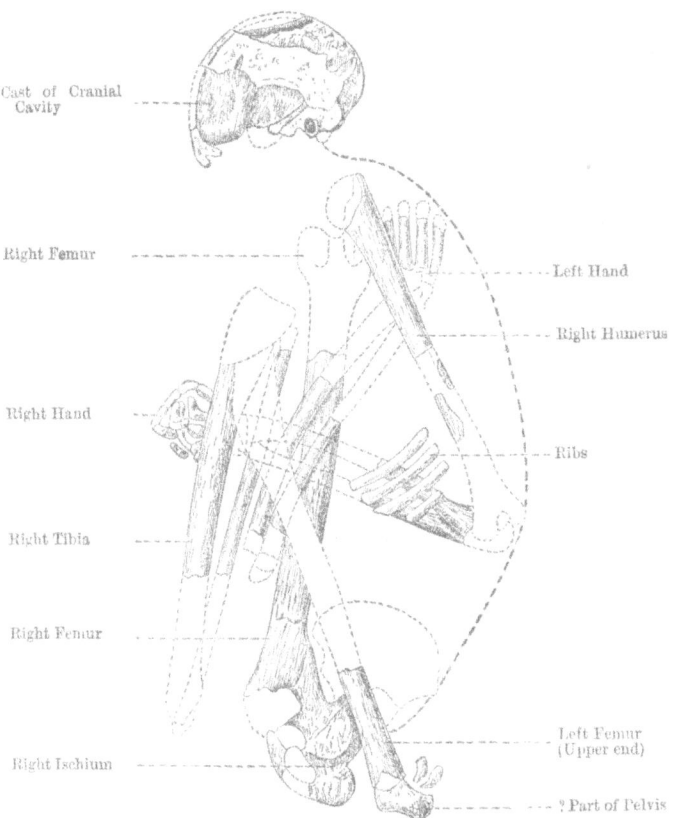

Fig. 29. Human skeleton found beneath Boulder-clay near Ipswich in 1911. (From the drawing prepared by Professor Keith, and published in the *East Anglian Daily Times*. Reproduced with permission.)

REFERENCES TO LITERATURE

CHAPTER I

1 Dubois, 1894. Pithecanthropus, ein Übergangsform, &c.
2 Blanckenhorn, 1910. Zeitschrift für Ethnologie. Band 42, S. 337.
3 Schwalbe, 1899. Zeitschrift für Morphologie und Anthropologie. From 1899 onwards.
4 Berry, 1910. Proceedings of the Royal Society of Edinburgh, XXXI. Part 1. 1910.
5 Cross, 1910. Proceedings of the Royal Society of Edinburgh, XXXI. Part 1. 1910.
6 Schoetensack, 1908. Der Unterkiefer des Homo heidelbergensis.
7 Keith, 1911. Lancet, March 18, 1911, abstract of the Hunterian Lectures.
8 Dubois, 1896. Anatomischer Anzeiger. Band XII, s. 15.

CHAPTERS II AND III

9 Avebury (Lubbock), 1868. International Congress for Prehistoric Archaeology.
10 Turner, 1864 (quoting Busk). Quarterly Journal of Science, Oct. 1864, p. 760.
11 Nehring, 1895. Zeitschrift für Ethnologie, 1895, s. 338.
12 Kramberger, 1899. Mittheilungen der anthropologischen Gesellschaft zu Wien. "Der Mensch von Krapina." Wiesbaden, 1906.
13 Marett, Archaeologia, 1911; also Keith, 1911. Nature, May 25, 1911. Keith and Knowles, Journal of Anatomy, 1911.
14 Boule, 1908. L'Anthropologie. Tome XIX. p. 519.
15 Klaatsch and Hauser, 1908. Archiv für Anthropologie. Band 35, 1909, p. 287.
16 Peyrony (and Capitan), 1909–1910. Bulletins de la Société d'Anthropologie de Paris, Jan. 20, 1910.

REFERENCES TO LITERATURE

17 Sollas, 1907. Philosophical Transactions of the Royal Society. Vol. 199 B.
18 Sera, 1909. Atti della Società romana di Antropologia, xv. fasc. II.
19 Verner, 1910. Ann. Rep. Hunterian Museum. R.C.S. London. Saturday Review, Sep. 16, 1911, and five following numbers.
20 Verneau, 1906. L'Anthropologie. Tome XVII.
21 Lehmann-Nitsche, 1907. Rivista del Museo de la Plata, XIV. 1907.
22 Lehmann-Nitsche, 1909. Naturwissenschaftliche Wochenschrift, Jena, VIII. 42.
23 Klaatsch, 1909. Prähistorische Zeitschrift, I.
24 Newton, 1895. Quarterly Journal of the Geological Society, August, 1895.
25 Schwalbe, 1906. "Der Schädel von Brüx." Zeitsch. für Morphologie und Anthropologie.
26 Hinton, 1910. Proceedings of the Geologists' Association. Vol. XXI. Part 10. 1910.

CHAPTER IV

27 Gaudry, 1888. Les ancêtres de nos animaux.
28 Schmidt, 1909. Archiv für Anthropologie. Band 35, s. 62, 1909.
29 Commont, 1908. L'Anthropologie. Tome XIX. p. 527.
30 Obermaier and Bayer, 1909. Korrespondenzblatt der Wiener anthropologischen Gesellschaft, XL. 9/12.
31 Rutot, 1900. Congrès international d'Archéologie préhistorique. Paris, 1900.
31 Rutot, 1904, ?1903. Quoted in Schwalbe 1906. "Vorgeschichte, usw." Zeitschrift für Morphologie und Anthropologie.
31 Rutot, 1911. Revue de l'Université. Brussels, 1911.

32 Penck, 1908. Zeitschrift für Ethnologie. Band XL. s. 390.
33 Laville, 1910. Bulletin de la Société d'Anthropologie de Paris, 1910.
34 Moir, 1910. Proceedings of the Geologists' Association, July 16, 1910. Prehistoric Society of East Anglia, 1911.
35 Warren, 1905. Journal of the Royal Anthropological Institute. Vol. XXXV., 1905, p. 337.
36 Boule, 1905. L'Anthropologie. Tome XVI. " Sur l'origine des Eolithes."
37 Obermaier, 1908. L'Anthropologie. Tome XIX. p. 613 (abstract), also p. 460 (abstract).
38 Grist, 1910. Journal of the Royal Anthropological Institute. Vol. XL. 1910, p. 192.
39 Sturge, 1909. Prehistoric Society of East Anglia, January 1909 (published in 1911).

CHAPTER V

40 Falconer, 1865. Collected Memoirs. Vol. II. p. 587.
41 Geikie, A. 1863. Text-book of Geology, 1903, p. 1312 and footnote *ibidem*.
42 Skertchley, 1878. The Fenland, p. 551.
43 Boule, 1888. Revue d'Anthropologie, "Essai de stratigraphie humaine."
44 Hoernes, 1903. Urgeschichte des Menschen. (2nd Edn, 1908.)
45 Obermaier, 1909. L'Anthropologie. Tome XX. p. 521.
46 Sollas 1908. Science Progress in the XXth Century, "Palaeolithic Man." (Reprinted in book-form, 1911.)
47 Boyd Dawkins, 1910. Huxley Lecture. Royal Anthropological Institute, 1911.

CHAPTER VI

48 Gaudry, 1878. Mammifères tertiaires.
49 Klaatsch, 1909. Prähistorische Zeitschrift. Band I.
50 Keith, 1911. Nature, Feb. 16, 1911...also Dec. 15, 1910.

REFERENCES TO LITERATURE

51 Schliz, 1909. Archiv für Anthropologie. Band 35, Ss. 239 et seq. "Die vorgeschichtlichen Schädeltypen der deutschen Länder."
52 Giuffrida-Ruggeri, 1910. Archivio per l'Antropologia e per la Etnologia, XL. 2.
53 Sera, 1910. Archivio per l'Antropologia e per la Etnologia, XL. fasc. 3/4.
54 Schwalbe, 1906. "Vorgeschichte des Menschen," Zeitschrift für Morphologie und Anthropologie.

Recent publications containing a summary of the latest discoveries.

Birkner. Beiträge zur Urgeschichte Bayerns. Bd XVII. 3/4. 1909.
Branco. Der Stand unserer Kenntnisse vom fossilen Menschen, 1910.
Buttel-Reepen. Aus dem Werdegang der Menschheit. 1911.
Giuffrida-Ruggeri. "Applicazioni, &c." Monitore Zoologico Italiano. No. 2. 1910. Rivista d'Italia. Agosto, 1911.
Keith. Hunterian Lectures, 1911. Ancient types of Mankind, 1911.
Kohlbrugge. Die morphologische Abstämmung des Menschen, 1908.
Lankester. The Kingdom of Man. 1906.
Leche. Der Mensch. 1911.
McCurdy. "The Antiquity of Man in Europe." Smithsonian Report (1909), p. 531. 1910.
Read and Smith, R. A. Guide to the Antiquities of the Stone Age. British Museum, 1911.
Rutot. Revue de l'Université. Bruxelles, January 1911.
Schwalbe. Darwin and Modern Science (Centenary volume), Cambridge, 1909.
Sollas. Palaeolithic Man. (Cf. No. 46 supra.) 1911.
Spulski. Zentralblatt für Zoologie. Band 17. Nos. 3/4. 1910.
Wright. Hunterian Lectures, Royal College of Surgeons, 1907.

INDEX

Acheulean type of implement, 83; *v. also* S. Acheul
Acromegaly, 141
Adloff, 30
Ameghino, 54, 80
Andalusia, 20, 76
Andaman islands, aborigines of, 49
Anthony, 37, 147
Anthropoid Ape (*v. also* Gorilla *and* Orang-utan), 3, 13, 14, 17, 22
Arctomys, 70, 73
Atlas vertebra, 53, 54
Aurignac, 49; implements of the type of, 70, 74, 81; skeleton from, 135-138, 145: v. also *Homo aurignacensis hauseri*
Australian aborigines, 50
Avebury, 17

Badger, 73
Baradero, 20, 53, 80
Bayer, 99
Berry, 9, 128
Bison *priscus*, 67; (species unknown), 72, 73, 75
Blanckenhorn (on Trinil strata), 4
Bos (? species), 72; *primigenius* (*v. also* Urus), 70, 74, 86, 139
Boulder-clay, 114, 115

Boule, 18, 20, 37, 45, 108, 109, 116, 117, 120
Brain, 3, 6, 7, 14, 37-39
Brain-case (as distinct from the face), 37, 45, 47, 55, 60-62
Branco, 54
Breuil, 108
Brow-ridges, 55, 61, 62
Brückner, 116
Brünn, 56, 57, 82
Brüx, 56, 57; strata, 81
Bury S. Edmunds, 134
Bush Race (South African aborigines), 50, 145
Busk, 19, 46

Canine fossa (of face), 36, 37, 55
Cave Bear, *v.* Ursus
Cave Hyaena, 78
Cervidae (*v. also* Stag), 67, 92
Chelles, implements, 68, 83, 98
Classification of human fossil remains, 60; also Table A
Combe-Capelle (Dordogne), 55, 56, 81
Commont, 98, 99, 105, 125
Corrèze (*v. also* La Chapelle), 71
Cranial base, 47
Croll, 116
Cro-Magnon, 79, 140
Cromer, forest-bed fauna, 66

INDEX

Cross, 9, 130-132 (diagram, p. 131)
Cyrena *fluminalis*, 83

Dawkins, Boyd, 125
de Bohun, château, 122
Dénise, 18, 147, 148
Dewlish, eoliths from, 109
Dolichocephalic proportions of skull, 55, 59
Dordogne, 20, 45: v. also *H. mousteriensis hauseri*
Duan, Eocene eoliths, 106
Dubois, references under *Pithecanthropus erectus*

Elephas *antiquus*, 66, 67, 70, 78, 87, 88-90, 101, 120; *meridionalis*, 101, 109; *primigenius*, v. Mammoth
Engis, 18, 19, 134, 147, 148
Eocene period, 106
Eoliths, 106-111
Erect attitude, 7, 61, 147

Falconer, 46, 114
Forbes Quarry (*v. also* Gibraltar), 19, 20, 32, 46-49, 76
Forest-bed, *v.* Cromer
Frizzi, 44

Galley Hill, 20; gravel pit, 82, 84; skeleton, 56-59, 86, 95, 130-132, 134
Gaudry, 50
Geikie, Sir A., 115
Geikie, J., 116
Germany, caves in, 95-98, 100
Ghilain, 109
Gibraltar (*v. also* Forbes Quarry), 19, 46-49, 76, 143-144
Giuffrida-Ruggeri, 140

Gorilla (*v.* Anthropoid Ape), 136-138
Grimaldi (*v. also* Grotte des Enfants), 50-52
Grotte des Enfants, 20, 76-79
Grotte du Prince, 120
Günz, glacial phase of, 119

Hauser, 39, 55: *v.* Homo
Heidelberg, v. *Homo heidelbergensis*
High-level terrace gravels (of Thames), 83
Hinton, 83, 101-104, 115, 125
Hippopotamus, 70, 78, 120
Hoernes, 20, 117, 120
Homo *aurignacensis hauseri*, 20, 55, 57, 135-138; *fossilis*, 20, 60; *heidelbergensis*, 1, 10-16, 22, 26, 27, 29, 32, 41-43; *mousteriensis hauseri*, 14, 20, 32, 39-45, 73; *neogaeus*, 20, 53-55; *primigenius*, 27, 60
Horse, 71, 73, 75
Huxley, 9, 135, 147, 148

Ibex, 73
Implements, sequence of, 102, 103
Interglacial phases, 67, 119, Table B
Ipswich skeleton, 148, 151-152

Jalón river (Aragon) implements, 101
Jawbone, 11-16, 26, 27, 29-31, 34, 37, 41-43, 53, 55, 60, 62
Jersey, *v.* S. Brélade
Julien, 116

Keith, 31, 137, 138, 140, 142, 144, 147

INDEX

Klaatsch, 20, 28, 36, 56; *diphyletic theory*, 135, 136, 139
Kramberger, 20, 24, 27, 30
Krapina, 20, 24–31, 32, 34, 42, 68–71; *fauna*, 91, 92

La Chapelle-aux-Saints, 20, 33–39, 47, 71
La Ferrassie, 20, 39, 45, 74, 75, 98
Laloy, 30
La Naulette, 18, and fig. 14
La Quina, preface, vi, 39, 150
Laville, 106, 125
Lehmann-Nitsche, 20, 54, 80
Le Mas d'Azil, 95, 97
Le Moustier, 29, 45; *cave*, 73–75: *v. also* Mousterian
Leontiasis *ossea*, 142
Levallois, 68
Limb bones, 50, 55
Löss, 79, 80; in Lower Austria, 124
Lyell, 114, 117, 147, 148

Macnamara, 46
Maffle, implements of, 83, 102, 104
Magdalenian period, 121
Malarnaud, 18
Mammoth, 18, 82, 92
Manouvrier, 15, 34, 38
Marett, 20, 30
Marmot, 70, 73
Mastoid process, 55
Mauer, v. also *H. heidelbergensis*, 65–66, 90, 104, 148
Mentone, *v.* Grimaldi *and* Grotte des Enfants
Mimomys, 88, 89
Mindel, glacial phase of, 119
Miocene period, 80
Moir, 106, 109

Monte Hermoso, 20, 53, **54, 80**
Morlot, 115
Mortillet, 117
Mousterian period, 121–125; *types of implement of*, 67, 68, 70, 71, 78, 94–98, 118, 134
Munck, 109
Mural decorative art in caves, 76

Neanderthal, 18, 19, 24, 27, 34–36, 38, 47, 55, 131–138, 147, 148
Negroid characters, 50, 52
Nehring, 20
Neolithic implements, 109
Newton, 20, 57
New World, *v.* S. America
Nicolle, 30
Northfleet, 57 : *v.* Galley Hill

Obermaier, 68, 99, 108, 116, 117
Ofnet, 96–98, 100
Oligocene period, implements in, 110
Orang-utan, 136–138: *v. also* Anthropoid Ape
Ostiaks, cranial form, 144

Pech de l'Aze, 20, 46, 75
Penck, 106, 107, 116–124, 126
Peyrony, 20, 45
Pithecanthropus erectus, 1–9, 14, 15, 31, 54, 63–65, 148
Pituitary gland and secretion, 141, 142
Pleistocene mammals and period, 66, 84
Pliocene strata, 64, 80
Prestwich, 114
Prince of Monaco, 50
Prognathism, 36, 50
Pruner-Bey, 49
Pygmy types of mankind, 49, 54

INDEX

Ramsay, 115
Reindeer, 71, 73–75, 78, 79, 86, 91, 92
Rhinoceros *etruscus*, 66, 87–89; *megarhinus*, 87–89; *merckii*, 67, 70, 78, 87, 89, 90, 92, 93, 96, 120; *tichorhinus*, 71, 73, 82, 92
Riss, glacial phase of, 119
River-drift, 115
Ronda, 49
Roth, 20
Rutot, 83, 102–107, 111

S. Acheul, 68, 101, 134
S. Brélade, 20, 30, 32, 71, 150, Table A
Saporta, 125
Schliz, 140
Schmidt, 95, 125
Schoetensack, 65, 66
Schwalbe, 4, 9, 20, 27, 46, 82
Scott, 80
Sera, 20, 46–48, 142–146
Sinel, 30
Sirgenstein, 96–98, 100
Skeletons, contracted position of, 73, 74, 78
Skertchley, 116, 117
Sollas, 20, 46, 124
Solutré-period and implements of, 124
South America, 20, 52–55, 79–81
Southern fauna, 67
Spy cave-men, 18, 19, 21, 24, 32, 34, 35, 44, 53
Stag, 75 : *v. also* Cervidae
Stature, 38, 44, 49, 59, 61
Steinmann, 80
Stone implements, value in evidence, 93

Strépy, implements of, 83, **102**, 104
Sturge, 109, 117
Suidae, *v.* Swine
Swine, 67, 92, 139

Taubach, 10, 20, 21–23, 31, 53, 67, 70, 86; *fauna*, 123; *implements*, 78, 98, 101
Teeth, 4, 10, 11, 14, 15, 21–23, 26, 27, 29–31, 41, 42, 50, 53, 60, 62
Tertiary mollusca, 80
Tetraprothomo, 54
Thames gravels, 83
Tilloux, implements and fauna of, 101
Tornqvist, 117
Trinil, 66 : v. also *P. erectus*
Trogontherium, 87, 89
Turner, 19

Ursus *arctos*, 70; *arvernensis*, 66, 88, 89; *deningeri*, 66; *spelaeus*, 66, 70, 72
Urus, v. *Bos primigenius*

Venezuela, 145
Verneau, 20, 50, 51
Verner, 20, 49
Voles, 92 : *v.* Mimomys

Walkhoff, 30
Warren, 108
Weiss, 67
Wildkirchli, 122
Wolf, 73
Würm : glacial phase of, 119
Württemburg, caverns of, 95–98, 100

For EU product safety concerns, contact us at Calle de José Abascal, 56–1°, 28003 Madrid, Spain or eugpsr@cambridge.org.

www.ingramcontent.com/pod-product-compliance
Ingram Content Group UK Ltd.
Pitfield, Milton Keynes, MK11 3LW, UK
UKHW040157230326
469255UK00012B/149